Citizen Action for Global Change

Syracuse Studies on Peace and Conflict Resolution

Harriet Hyman Alonso, Charles Chatfield, *and* Louis Kriesberg
Series Editors

Four Neptune Group leaders at Geneva in 1978. From left: Sister Mary Beth Reissen, Miriam Levering, Sam Levering (background), and Lee Kimball. Photograph courtesy of Lee Kimball.

CITIZEN ACTION
for
GLOBAL CHANGE
The Neptune Group and Law of the Sea

Ralph B. Levering *and* Miriam L. Levering

Syracuse University Press

Copyright © 1999 by Syracuse University Press
Syracuse, New York 13244-5160
All Rights Reserved

First Edition 1999

99 00 01 02 03 04   6 5 4 3 2 1

The paper used in this publication meets the minimum requirements of American National Standard for Information Sciences—Permanence of Paper for Printed Library Materials, ANSI Z39.48—1984. ∞™

Library of Congress Cataloging-in-Publication Data
Levering, Ralph B.
 Citizen action for global change : the neptune group and law of the sea / Ralph B. Levering and Miriam L. Levering.
  p. cm. — (Syracuse studies on peace and conflict resolution)
 Includes bibliographical references and index.
 ISBN 0-8156-2794-7 (cloth : alk. paper). — ISBN 0-8156-2795-5 (pbk. : alk. paper)
 1. Law of the sea. 2. United Nations Convention on the Law of the Sea (1982)   I. Levering, Miriam L., 1913– . II. Title.
 III. Series.
 KZA1120.3.L48   1999
 341.4'5'—DC21   98-45026

Manufactured in the United States of America

For George and Marilyn White, and for everyone else who participated in the Neptune Group and related NGOs

Ralph B. Levering is professor of history at Davidson College in North Carolina. He is the author of *American Opinion and the Russian Alliance, 1939–1945* (1976), *The Public and American Foreign Policy, 1918–1978* (1978), and *The Cold War: A Post–Cold War History* (1994). He is co-author of *The Kennedy Crises: The Press, the Presidency, and Foreign Policy* (1983).

Miriam L. Levering was a peace and world-order activist from *the 1930s until her death in 1991. She a*lso was a prominent Quaker. She is the author of *Love, Mom: Stories from the Life of a Global Activist, Teacher and Mother of Six* (1996).

Contents

Illustrations	ix
Preface	xi
Abbreviations	xv
Introduction	xvii
1. Setting the Stage	3
2. Getting Started and Mobilizing Influence	19
3. Opposing Unilateralism in Washington	35
4. Gaining Legitimacy at the United Nations Conference	51
5. Helping to Shape the Treaty	72
6. Winning and Losing on Capitol Hill	106
7. Seeking to Counteract the Reagan Shocks	128
8. Assessing and Learning from the Neptune Group	149
Appendix: Meetings of UNCLOS III	169
References	171
Index	179

Illustrations

Neptune Group leaders at Geneva, 1978	*frontispiece*
Manganese nodules from the ocean floor	22
Part of the OEP team at Caracas, 1974	54
At Geneva in 1975	62
Leaders of the ecumenical prayer service in New York, 1977	88
Bernardo Zuleta, Hamilton Amerasinghe, and David Hall	90
Sam Levering and Adolf Schneider at Geneva, 1978	92
Tommy Koh	99
Miriam Levering at OEP office, early 1980s	133
Signing at Montego Bay, 1982	147

Preface

Because it involves a mother-son collaboration, research and writing that lasted even longer than the record-setting nine years of the conference being studied, and two very different manuscripts (hers from the mid-1980s and mine written a decade later), how this book—the first one about a particular NGO's work at a UN conference—came to be is itself a story worth telling.

My mother, Miriam Levering, a leading NGO (nongovernmental organization) representative at the Third United Nations Conference on the Law of the Sea (UNCLOS III), began making plans to write this book in the fall of 1981, more than a year before the conference formally ended with the signing ceremony at Montego Bay, Jamaica, in December 1982. Miriam knew that the members of the NGOs with which she had worked at the conference for nearly a decade, the so-called Neptune Group, had made important contributions to the negotiations. She wanted to tell their story not only to highlight the work of the conference and thus build support for the new ocean treaty, but also to enable NGO representatives and diplomats working in future international negotiations to learn from the Neptune Group's experiences. A longtime activist on behalf of world order, Miriam also wanted to inspire others to devote their lives—as she had hers—to strengthening the international laws and institutions that, in her view, offered the best hope for a more just and peaceful future.

At the time of her death at the age of seventy-eight in October 1991, Miriam remained as determined as ever to help bring the 1982 treaty to fruition. Had she lived, she would have been thrilled when, after serious negotiations in 1993 and early 1994, the United States signed an amended treaty in July 1994 that became international law that November. Although the Clinton administration has honored the terms of the as yet unratified treaty, she would have kept working to urge the Senate to ratify it and to build support for the new law of the sea and for international governance more generally.

A diplomatic historian who had written several books on U.S. public opinion and foreign policy and on the cold war, I offered in late 1988 to help write the book on the Neptune Group's experiences. During my sabbatical in academic 1989-90, Miriam set up and helped me conduct fifty-one interviews—mainly in Washington, New York, and Boston. During that year I also read the relevant secondary literature and organized and took notes on the couple thousand letters, the couple hundred Neptune Group publications, and the other materials (including numerous contemporary clippings from newspapers and articles in periodicals) in about twenty-five large boxes that Miriam turned over to me.[1] The book took several more years to complete; my only regret is that Miriam and my father, Sam Levering, an NGO activist equally committed to law of the sea who died in December 1994, did not live to see it.

Following the inclinations of its authors, the book combines two literary styles, history and memoir. I wrote the introduction, chapter 1, and the conclusion (chap. 8); these sections feature methods of analysis common in history and in other social sciences. In contrast, the main narrative section of the book, chapters 2-7, combines history and memoir. This section qualifies as memoir in three main senses. First, in composing these chapters—especially 2 and 3—I drew heavily on Miriam's earlier manuscript. Because Miriam was an engaging writer with a gift for memorable phrases, something like 15 percent of the words in chapters 2–7—and a larger percentage of the interpretations—were incorporated from Miriam's manuscript. Second, I had numerous long conversations with my parents about the Neptune Group's work throughout academic 1989–90 and in subsequent years; I recorded several of these conversations, and used notes from them in writing these chapters.

Third and most important, I decided early on to present the material in the narrative section largely through what I understood to be my mother's and father's perspectives, rather than that of an outside

1. In 1998, these boxes of materials were added to the Levering family papers in the Friends Historical Collection of Guilford College in Greensboro, North Carolina. Sources from these materials are cited in the references (pp. 171–77) as being located in ML files.

observer trying to weigh and balance diverse perspectives on the Neptune Group's work. I did so because Miriam wanted to tell the Neptune Group's story as she saw it, because I came to see that such an approach provided focus and integrity for the argument, and because I did not want to feel compelled, as a historian, to question at every opportunity things that my parents said or did that appeared inconsistent or overly partisan in retrospect.

Following the form used in Mother's manuscript, I have told the basic story (chapters 2–7) in the first person, using "I" to refer to Miriam's views and "we" to refer to Miriam's and Sam's views and, often, to the views of their close associates as well. In these chapters, therefore, I have sought to tell Miriam's and Sam's story from their perspectives, supplemented by research in the written record and by insights from the people we interviewed. The six central chapters thus might be called a jointly written history/memoir that one of the authors was unable to review because of her death.

Detailed comments by one of Miriam's closest friends and associates in the Neptune Group, Lee Kimball, alerted me to at least one instance in these chapters where the emphases I was presenting may have differed somewhat from Miriam's views at the time. In revising the manuscript for publication, I responded conscientiously to Kimball's suggestions. Even so, however, these chapters must necessarily be viewed both as a work of history and a participant's memoir.

On behalf of both Miriam and myself, I wish to thank all those who helped us with this project, including the fifty-eight people who were interviewed by Miriam and me, by me alone, or by Miriam's able Neptune Group associate, Eleanor Smith. The fact that so many accomplished and busy people agreed to what were often lengthy interviews was itself a tribute to the Neptune Group's work during UNCLOS III—and to Miriam and Eleanor personally. Four of the interviewees—Robert Cory, Joanna Esterly, Martin Glassner, and Jessica Mott—kindly provided lodging and meals for Miriam and me during our research trips in 1989 and 1990. So did two of my sisters, Montague Kern and Betsy Morgan, and two cousins, David Lindsey and Phil Wellons. Marie Wilkinson of the Council on Ocean Law (formerly Citizens for Ocean Law) helped us update addresses and phone numbers.

I also wish to thank the officials, archivists, and librarians who helped us find important documents: Sharon Byrd and Gina Overcash of the

Davidson College library; Scott Maslow of the State Department's Office of Freedom of Information, Privacy, and Classification Review; David Patterson of the State Department's Historian's Office; Mark Shenise of the Methodist Church's General Commission on Archives and History; Steven Tilley of the National Security Council; and Marsha Trimble of the University of Virginia Law Library. This library currently holds the premier collection of documents relating to UNCLOS III.

Those who read and commented on the manuscript provided invaluable assistance: Charles Chatfield, Calvin Davis, Earl Edmondson, Alan Henrikson, Gary Hess, Lee Kimball, Frank Levering, Alton and Elizabeth Lindsey, Arthur Paterson, and Jack Perry. Growing less stubborn with each passing year, I made most of the changes they suggested!

Davidson College provided an ideal atmosphere for research and writing. Two colleagues were especially helpful: Dean Robert Williams, who granted a sabbatical in 1989-90 and a leave in spring 1994, and Kristi Newton, the history department's cheerful, capable secretary. The entire history department was highly supportive. Three of Davidson's librarians—Jan Blodgett, Molly Gillespie, and Jason Hamrick—patiently helped me communicate with my computer.

My five siblings and many other family members and friends repeatedly offered encouragement. I am especially grateful to my wife, Patricia Webb Levering, without whose loving support and wise counsel this book would never have been completed.

This book is dedicated to the many staff members and volunteers who worked in the Neptune Group organizations and associated NGOs, and to all who donated money, lodging, or meals to support the cause. In particular, the book is dedicated to George and Marilyn White of Charlotte, North Carolina, longtime associates of the Leverings in the American Freedom Association. The Whites epitomize the selfless dedication to advancing global governance and understanding of so many people, most of whom are not mentioned by name in these pages. Whether named or unnamed, all of you have helped make history.

Abbreviations

AFA	American Freedom Association
FCNL	Friends [Quaker] Committee on National Legislation
G11	group of mostly developed nations at UNCLOS III
G77	group of 77 mostly developing nations at UNCLOS III
INGO	international nongovernmental organization
ISA	International Seabed Authority, often called "the Authority"
ML files	Miriam Levering's manuscript collection of materials relating to UNCLOS III
NGO	nongovernmental organization
OEP	Ocean Education Project of the American Freedom Association
Public Advisory Committee	Public Advisory Committee to the Joint Interagency Task Force on Law of the Sea
SAIS	School of Advanced International Studies, Johns Hopkins University
UMLSP	United Methodist Law of the Sea Project
UNCLOS III	Third UN Conference on Law of the Sea
U.S. Committee	United States Committee for the Oceans

Introduction

At the last meeting of UNCLOS III in 1982, Martin Glassner, a frequent attender of the conference and a professor of geography at Southern Connecticut State University, told Miriam Levering that she absolutely had to write a book about the Neptune Group's work. "I was so impressed with the Neptune Group—the quality of the work and the dedication of the individuals—that I thought that a book should be written about the group," Glassner recalled (Glassner 1990). He wanted to assign it to students in his world political geography classes, he said, to inspire them to realize what a few ordinary people could do to make the world better. The most important problems facing the human race—for example, damage to the environment, population pressures, the gap between rich and poor, the control of weapons of mass destruction—were now global, Glassner believed, and ordinary people needed to feel empowered to work to find solutions to them.

Like Glassner, Miriam was amazed by what she and the relatively few world-order activists who worked with her had accomplished both at the conference and in Washington. A longtime peace activist who believed that education in the broadest sense was the key to human betterment, Miriam also shared the warmhearted professor's desire to spread the word about what the Neptune Group had accomplished.

Fortunately for the Leverings and their associates, who normally put in long hours for modest salaries or for no pay and who occasionally felt discouraged by the frequent deadlocks at the conference and the growing skepticism about it in Washington, Neptune Group representatives repeatedly felt encouraged by praise for their work. In September 1978, for example, Sen. Charles H. Percy of Illinois, a leading Republican on the Senate Foreign Relations Committee, praised Sam Levering and Barbara Weaver of the United Methodist Law of the Sea Project as follows: "Your own efforts to achieve a comprehensive, widely-accepted treaty are commendable. If we are able to develop a regime which the U.S. Senate can accept, it will owe a great deal to your careful and informed work" (Percy 1978)

The biggest thrill for the Leverings and their colleagues occurred at the signing session for the treaty in Montego Bay, Jamaica, in December 1982. In the presidential address Tommy T. B. Koh of Singapore praised the Neptune Group as the NGO representatives who had contributed the most to the conference. Koh did not mention any other NGOs. After discussing eight other "features of the negotiating process . . . which were productive," Koh offered his ninth point:

> I should also acknowledge the role played by the non-governmental organizations, such as the Neptune Group. They provided the Conference with three valuable services. They brought independent experts to meet with delegations, thus enabling us to have an independent source of information on technical issues. They assisted representatives from developing countries to narrow the technical gap between them and their counterparts from developed countries. They also provided us with opportunities to meet, away from the Conference, in a more relaxed atmosphere, to discuss some of the most difficult issues confronted by the Conference. (Koh 1982, 3, 4–5)

In short, Glassner was right: the Neptune Group's story needed telling, both because of its importance in the larger story of UNCLOS III and because of the inspiration—and lessons—that ordinary citizens and other activists seeking increased world order and a cleaner environment can draw from the Neptune Group's experience.

A natural storyteller who enlivened her speeches on strengthening global governance with engaging anecdotes, Miriam also knew that she had a good story. How had it happened that she, a former high school teacher, and her husband, Sam, an orchardist who continued to operate their farm in southwestern Virginia even as he spent most of his time working on law of the sea, had become respected, influential activists on the issue? Miriam and Sam had neither legal training nor personal financial resources. Yet Sam headed the subcommittee on the ocean environment of the U.S. State Department's Public Advisory Committee on Law of the Sea and took part in frequent discussions with policy makers at the State Department and with legislators on Capitol Hill. And Miriam and her colleagues offered a variety of services free of charge during and between the sessions of UNCLOS III that were highly valued by leaders of the conference and by ordinary delegates alike.

This book's most important purpose is to elucidate the group's work at the conference, thus suggesting what NGOs in general may be able to contribute to international negotiations in the future. Another central goal is to explain and evaluate the Neptune Group's work away from the conference, especially its efforts in Washington to encourage the four administrations that governed from 1972 through 1982 to pursue broad-minded, conciliatory approaches to the negotiations and to prevent—or at least to postpone—congressional passage of the unilateral seabed mining bills that well-funded business interests and their allies on Capitol Hill pushed throughout the 1970s.

The group's work at UNCLOS III, its activity in Washington, and its efforts to build public support for the negotiations were interrelated and increased its effectiveness, especially at the conference. This increased effectiveness at UNCLOS III largely resulted from the appreciation of many delegates, especially from developing countries, for the group's efforts to prevent congressional passage of the unilateral seabed mining bills. Many delegates viewed this proposed legislation as an arrogant, unjust effort to control resources that belonged to the entire human community, not to the nation or small group of nations that had the largest navies and the most sophisticated mining technology.

This book thus examines the group's work in three main areas, or arenas: the conference, the administration, and Congress. Its work in a fourth arena, the American public (including its work with the main channel for ideas flowing to the public, the news media), receives far less attention. Given the absence of public opinion polls or elections contested on this issue, the results of the group's efforts to influence public-opinion are the most difficult to gauge.

This four-arena scope was unique among U.S.-based NGOs working on law of the sea at the time and, as suggested above, contributed to the Neptune Group's effectiveness relative to other NGOs. Also contributing to the group's relative effectiveness was an attribute emphasized by Elliot Richardson, who headed the U.S. delegation from 1977 through 1980. "There wasn't another NGO I can think of," Richardson recalled, "that existed simply in the interest of the treaty" (Richardson 1990).

To provide additional background information for succeeding chapters, the first chapter offers concise answers to three sets of questions. First, why did UNCLOS III take place? Why did the conference last so long? And what did it accomplish? Second, what was the Neptune Group? Why did it arise, and what were its goals? And, to return to two key protagonists, who were Sam and Miriam Levering? Third, where does this study fit into the scholarship on NGO involvement in international negotiations? Readers knowledgeable about UNCLOS III might want to skip the first section; readers uninterested in social science theory—even if clearly discussed—might want to jump past the third.

A fourth important set of questions—in general, what other NGOs were involved in the conference, and how effective were they?—is addressed toward the end of chapter 2. Further references to other NGOs with which the Neptune Group worked in the United States and at the conference occur throughout the book. The reader should bear in mind that this is a history/memoir of the Neptune Group's work, not a comprehensive study of NGO involvement in UNCLOS III.[1] Miriam's firsthand experience and our subsequent research suggest that activists in NGOs such as the Neptune Group and the Sierra Club and employees of businesses such as Shell and Lockheed had the greatest impact on the conference of all the nongovernmental representatives.

---

1. The only comprehensive study to date is a doctoral dissertation by John Edward Lacey (1982). While useful in some ways, this study suffers from excessive quantification and from an analytical approach that avoids evaluating contributions by particular individuals and NGOs. As Lacey points out (309–11), the UN secretariat accredited to attend UNCLOS III only representatives of international NGOs (INGOs)—that is, ones with branches in more than one country. Thus Miriam Levering of OEP and Sam Levering of the U.S. Committee for the Oceans received their accreditation through the Friends (Quaker) World Committee for Consultation, and other Neptune Group members were accredited by other INGOs.

Citizen Action for Global Change

"The most important political development of the second millennium was the firm establishment, first in one or two countries, then in many, of the rule of law. . . . What we have seen in the last century, first with the League of Nations, then with the United Nations (and its associated bodies) is the attempt to create a global rule of law. Like its predecessor—the rule of law in individual states—getting it in place and working is going to be a long, difficult and occasionally bloody struggle. . . . We will surely get there eventually, and the rule of law throughout our planet is likely to be among the achievements of the third millennium, as its establishment nationally was of the second."

—British historian Paul Johnson,
Wall Street Journal, March 10, 1999

1

Setting the Stage

UNCLOS III TOOK PLACE, in Clyde Sanger's apt phrase, because "[m]ost countries had something important they wanted to settle, or change" (Sanger 1987, 3). By the late 1960s officials in many nations had concluded that traditional law of the sea was obsolete, that changes in technology and thus in the ability to exploit resources (fish, oil, minerals, etc.) on a large scale and to conduct sophisticated military and scientific operations under the ocean's surface beyond the customary three-mile territorial limit necessitated the development of what Columbia University professor Guilio Pontecorvo has called a "new order of the oceans" (Pontecorvo 1986).

Developed nations, notably the United States and the Soviet Union, were concerned that unilateral claims to ever-larger offshore territorial and resource zones would hamper the global operations of their navies, their air forces, and their fishing fleets. Developing nations, aware by 1967 that the deep seabed contained large quantities of nickel and other minerals, feared that ships from developed nations would harvest these minerals and keep the profits, thus widening the gap between rich and poor. Moreover, the era's "cod wars" between Great Britain and Iceland and the "tuna wars" involving the seizure of U.S. fishing vessels off the west coast of South America grew out of nations' conflicting claims in regard to territorial limits and ocean resources. An updated law of the sea clearly was needed.

The first two UN conferences on law of the sea (UNCLOS I and II), held in 1958 and 1960 with sharply limited agendas, failed to resolve the increasingly urgent problems of ocean management. Moreover, many of the newly independent nations of the post-1945 era did not endorse the limited agreements reached at these conferences, largely because they had not participated in them. If a widely accepted ocean treaty was to be fashioned, therefore, the UN (a) would have to invite all nations to take part and (b) would have to ask the conference to consider the full range of issues concerning claims to territory and control of resources. Such an approach would ensure that all nations could potentially benefit, materially as well as legally, from a comprehensive treaty. Despite the preference of some developed nations for a narrower agenda, the 1970 UN General Assembly resolution that established UNCLOS III reflected the desire of the developing nations, a majority of UN members, for universality and comprehensiveness.[1]

Why did UNCLOS III last nine years (December 1973–December 1982), much longer than any previous international conference? There were many reasons, including the ambitious agenda and the fact that it took years for many delegates from newly independent nations to learn both the details and the process of the negotiations. "On top of the lack of factual information and analysis," Lee Kimball of the Neptune Group recalled, "many delegates were unfamiliar with international negotiation as a process, and even less so with the wiles of crafty drafters of treaty language" (Kimball 1995).

The negotiations also lasted for so long because leaders of developed nations—notably the United States—found it virtually impossible to prioritize their interests at the conference. The Department of Defense, for example, was most concerned about issues of navigation and overflight, whereas the Department of the Interior emphasized assured access by U.S.-owned companies to minerals on and beneath the seabed. Although aware of these and other specific interests, many State Department officials—like many in the Neptune Group—believed that the overall U.S. interest in achieving a new, fully elaborated law of the sea was greater than any particular interest such as seabed mining. In practice, U.S. and other Western negotiators contributed to the length of the negotiations by refusing to trade away one set of interests (for example,

---

1. Markus G. Schmidt, the most thorough scholar of the negotiations to date, has noted that many developed nations would have preferred a more limited agenda focused on the "question of maximum breadth of the territorial sea and the related straits passage issue . . . but the developing countries, their number then swelled by many newly independent states, refused to consider these questions in isolation from other issues, mostly related to resources" (Schmidt 1989, 22).

on deep seabed mining) to achieve their goals on another set (for example, navigation).

Although these factors are important, the primary reason for the lengthy proceedings was the complicated, seemingly endless negotiations over how and under what legal and administrative controls the mining of hard minerals on the deep seabed beyond national jurisdiction should be allowed to proceed. These negotiations—often pitting delegates from the "Group of 77" (G77) developing nations against representatives from the United States and other developed nations—almost certainly caused the conference to drag on for several years longer than it otherwise would have. Divisions on deep seabed mining within the G77 and within the industrialized nations also contributed to the length and complexity of the negotiations.[2] Because deep seabed mining was a prospective rather than an existing industry, it was extremely difficult to fashion detailed rules and regulations, to agree on probable costs and profits, and to determine what percentage of the revenues, if any, should be used to cover administrative costs under the treaty or be distributed to developing nations.

These problems, in themselves, would have made negotiations on deep seabed mining difficult. In fact, the negotiations became almost impossible largely because of sharply contrasting ideological and political perspectives that divided many G77 delegates from most delegates from the developed countries. Many G77 delegates were caught up in the then-fashionable enthusiasm for state-owned enterprises in developing countries and the excitement about calls for a "new international economic order" in which richer nations would be

---

2. The G77 delegates tended to divide into moderates who wanted to make some concessions to Western negotiators, and hard-liners who opposed major compromises. The delegates from the developed nations also differed among themselves on many issues. Canada, for example, with its large supplies of nickel and other hard minerals, supported G77 demands for stringent limits on production of hard minerals from the deep seabed, whereas the United States, an importer of nickel and other hard minerals, opposed such limits. Representatives of major western European nations often took a "moderate" position between the United States and the G77, though opposition among western European business leaders and delegates to the G77 positions on seabed mining became increasingly evident in the late 1970s and early 1980s. By far the best study of the negotiations on seabed mining is Schmidt 1989.

forced to help poorer ones catch up economically. These delegates insisted that a public international agency (the "Authority") effectively controlled by representatives of developing nations be set up to regulate seabed mining, that a public international corporation (the "Enterprise") controlled by the Authority conduct much or all of the actual mining, and that developed nations provide most of the funding for both of these agencies. Moreover, many G77 delegates insisted that, to ensure the Enterprise's success, the treaty should require Western companies to share with the Enterprise their seabed mining technology and expertise.

To put it mildly, many of these G77 proposals—especially the one requiring transfers of technology that appeared to nullify existing laws relating to patents and, in general, to challenge private ownership of technology—were not popular were business leaders and politicians in the major industrialized countries. But because G77 delegates held the upper hand in the committee writing Part XI, the section of the treaty dealing with deep seabed mining, many of these proposals found their way into the negotiating text. The main task of U.S. and other Western negotiators from 1976 forward, therefore, was to try to make Part XI more congruent with their own nations' laws and business practices. The effort to make Part XI acceptable to Western business executives and politicians was still going on when the final negotiating session ended in April 1982.

During the late 1970s and early 1980s, the heated debates about Part XI in the negotiations and in national business and political forums often overshadowed the conference's many achievements. In retrospect it can be seen that, through an arduous building of consensus and not permitting votes until the treaty was completed, delegates from more than 150 countries constructed a body of ocean law that, in its essential outlines, is likely to last for a long time. Among other things, the treaty established uniform twelve-mile territorial seas and two-hundred-mile economic resource zones, enumerated detailed rules relating to passage through straits and archipelagoes, spelled out the maritime rights of landlocked and other "geographically disadvantaged" states, defined "continental margin" and delineated coastal state jurisdiction over ocean resources, increased protection for the marine environment, and set up mechanisms for the peaceful settlement of disputes.

By the mid-1980s the world community, including the United States, had embraced as customary international law all of the treaty's provisions except those relating to deep seabed mining.³ By November 1994, when the treaty with a revised Part XI took effect, the negotiating process that the UN General Assembly had set in motion twenty-four years earlier was finally complete. Although some of his listeners may have considered his judgment premature and exaggerated when he offered it during a meeting of the U.S. Public Advisory Committee on the Law of the Sea in November 1980, State Department official Carlyle Maw's bold assessment has stood up: "What has been accomplished [at UNCLOS III] is truly remarkable—the greatest accomplishment in the development of international law of all time" (Public Advisory Committee 1980).

The Neptune Group

The Neptune Group was not an organization. Instead, it was a name that delegates and other attenders at the sessions of UNCLOS III gave to a constantly changing collection of NGO representatives who sponsored programs and receptions at the conference and who jointly published a newspaper, *Neptune*, beginning at the session in Geneva in the spring of 1975 and continuing through sessions in the early 1980s. The leaders of these NGO representatives at the conference—and hence of the so-called Neptune Group—were Miriam Levering of the Ocean Education Project (OEP), founded in January 1973, and Lee Kimball and Barbara Weaver of the United Methodist Law of the Sea Project (UMLSP), begun exactly two years later. In addition, the United States Committee for the Oceans (hereinafter called the U.S. Committee), the Washington lobbying organization founded in early 1972 and headed by Miriam's husband, Sam Levering, was integral to the Neptune Group because Sam often attended

---

3. Tom Kitsos, a congressional staffer who had frequent contacts with Sam Levering and other Neptune Group members in regard to deep seabed mining legislation in the late 1970s and early 1980s, told us that congressional staffs and diplomats in Washington viewed the treaty—except for the section on deep seabed mining—as customary international law by the mid-1980s (Kitsos 1990). Elliot Richardson, the head of the U.S. delegation from 1977 through 1980, echoed this view: "Nearly all of its [the treaty's] provisions for the use, management, protection, exploration, and exploitation of the oceans and their resources have already been assimilated into the body of customary international law" (Richardson, "Foreword," in Schmidt 1989, 5).

conference sessions and consulted with leaders of UNCLOS III in developing his strategy for opposing unilateral seabed mining bills on Capitol Hill. Thus the Neptune Group comprised three organizations, two of which made important contributions at the conference and all of which were heavily involved in building support for ocean law in Washington and throughout the United States.

To note that three organizations formed the Neptune Group, while strictly accurate, is to miss an important point: the Leverings and their associates often networked with representatives of other groups to stretch scarce resources in setting up programs or to build a broader base of support for their ideas. At the conference, for example, the two Neptune Group organizations frequently cosponsored programs with such groups as the Sierra Club, the World Affairs Council of Philadelphia, the Stanley Foundation, the Quaker UN Program, and Quaker House in Geneva. In Washington, cosponsors included such groups as Members of Congress for Peace Through Law, the American Friends Service Committee, Arlie House, and the School of Advanced International Studies at Johns Hopkins University (SAIS). Because representatives of these and other NGOs, foundations, and religious and educational institutions often attended meetings that OEP and UMLSP had largely organized, it is not surprising that delegates and UN officials often did not know exactly who—apart from the Leverings, Kimball, and Weaver—belonged to the Neptune Group.

Why did the Neptune Group arise, and what were its goals? Broadly speaking, the group arose because, throughout the twentieth century, there have been a substantial number of Americans of vision and resources (with active but less numerous counterparts in western Europe and elsewhere) who have believed that the United States and other democratic nations should take the lead in building a world ruled by law and orderly government to replace the international anarchy that led, in their judgment, to the two devastating world wars between 1914 and 1945 and to numerous smaller wars thereafter. Historians have labeled these members of the broader U.S. peace movement who supported greater world governance "internationalists" or "liberal internationalists," partly to distinguish them from the pacifists who made up the movement's other major wing.[4]

---

4. Good summaries of the growing splits between pacifists and internationalists in the early twentieth century are contained in Chatfield 1992, 43–50, and DeBenedetti 1980, 79–119.

Liberal internationalists in the United States helped conceptualize and advocate the League of Nations during and after World War I, and they helped build public support for the United Nations during and after World War II in such NGOs as the United Nations Association. Those who supported more ambitious moves toward world government—including the Leverings and the two Philadelphians who provided the initial funding for their work, William Fischer and A. Barton Lewis—joined such organizations as the United World Federalists, which the Leverings helped found in 1947. Since the early 1900s, liberal internationalism also has been an important influence in numerous Protestant denominations, including the Leverings' Quakerism and Weaver's Methodism. The Neptune Group thus grew out of both the secular and religious components of the U.S. peace movement's liberal internationalist wing.

The commitment to strengthening world governance that the Leverings and many of their colleagues and financial backers felt was every bit as strong and durable as the commitment that activists in other movements—for example, civil rights, women's rights, and the environmental movement—have felt and acted upon. Activists for world governance believed that they could change the world.

Drawing upon the analogy of the weak Articles of Confederation being transformed into the strong yet limited U.S. Constitution in 1787, the Leverings and other world federalists hoped in the 1940s and 1950s that the weak United Nations could be changed through international agreement into an organization powerful enough to prevent arms buildups and wars, yet limited enough to permit self-government in most areas of life for the world's diverse peoples.[5] By the late 1960s, however, the Leverings and many other world federalists, including Fischer and Lewis, were aware that the leaders of most nation-states strongly opposed giving the UN supreme power in the military area. They thus believed that the goal of increased world governance could best be advanced by working in areas such as law of the sea that were less subject to appeals to compatriots based on fear and national pride than military security was.

---

5. On the world government movement in the United States during and after World War II, including the United World Federalists, see Baratta 1987 and Wooley 1988, 3–82. A fine bibliography, including many works that the Leverings and other World Federalists read over the years, is Baratta 1987.

Lee Kimball, who worked for OEP in late 1974 and early 1975, succinctly stated the paramount early goal of the Leverings' organizations in a letter written at the time to a contact in England: "The Ocean Education Project is a non-governmental group interested in seeing an effective international organization result from these Law of the Sea Conferences capable of implementing the concept of the oceans and the deep seabed beyond national jurisdiction as the 'common heritage of mankind' " (Kimball 1975a).

Although the members of the Neptune Group never abandoned the ambitious goal of substantially strengthening international institutions, over time the more modest goal of reaching agreement on a treaty and encouraging U.S. leadership in implementing it took precedence in both OEP and UMLSP publications. And while the UMLSP supported increased world law and governance, the leaders of the Methodist project, aware of the U.S. public's strong support in the 1970s for environmental protection, placed more emphasis in their publications and speeches than the Leverings did in theirs on the conference's efforts to protect the ocean environment. UMLSP stressed that protection of the oceans—especially their fish, mammals, and other living resources—was proper stewardship of God's creation.[6]

And who were the Leverings?[7] Nearly sixty-four years old when he began full-time work on law of the sea early in 1972, Sam was the son of Ohio Quakers who in 1907 started the orchard in southwestern Virginia where Sam grew up and where, after working for five years in New Deal Washington, he returned in 1939 to take over the business from his father. While a graduate student in fruit growing at Cornell University in the early 1930s, Sam decided to devote his life—apart from the time needed to earn a modest living—to working for world peace.

Between 1939 and 1972, Sam, who normally considered leisure a waste of his precious time, devoted nearly half of his waking hours to the orchard and most of the other half to working for peace through

---

6. For examples of the emphasis on environmental protection in UMLSP publications, see *Whales: How Safe the Gentle Giants of the Sea?* (1978) and *Ten Commandments of the New Earth* (1979).

7. The Leverings' major associates in the Neptune Group, including Lee Kimball and Barbara Weaver, will be discussed in chapter 2. Partly owing to modesty, Miriam Levering did not describe Sam Levering or herself in detail in her original manuscript, and I have followed her example in chapters 2–7.

both religious and secular organizations. He took leading policy-making roles in several Quaker organizations, notably the Friends Committee on National Legislation in Washington and Friends United Meeting in Richmond, Indiana. A highlight of Sam's church work was heading a small group of Quaker leaders who met with President John Kennedy in May 1962 to urge improved relations with communist China.

In the secular arena, Sam was a founder of the United World Federalists in 1947 and held high positions in the organization at midcentury. He also spoke frequently across the country for world federalism and against the nuclear arms race. On one occasion he debated the merits of world federalism with well-known political scientist Hans Morgenthau, and on another he challenged U.S. reliance on nuclear weapons in a debate with renowned atomic scientist Edward Teller. False—but still harmful—accusations by right-wing activists during the McCarthy era that world federalists were communists or communist sympathizers prompted southeastern federalists to change the name of their major regional organization to American Freedom Association (AFA). Both Sam and Miriam were leaders in AFA from the 1950s through the 1980s, and Sam also played a key role in the late 1960s in a North Carolina group that opposed the Vietnam War.

Sam was well-known within Quaker and peace circles not only because he was an informed and inspiring speaker, but also because he exhibited great drive and determination. "If anything is worth doing," he often said, "it is worth doing well." The same qualities that earned him a straight-A average as an undergraduate at Cornell before the era of grade inflation helped him master the available information about—and the congressional politics of—deep seabed mining in the 1970s. Like Lee Kimball, who also took justifiable pride in her expertise, Sam was more knowledgeable about the congressional bills and the negotiating texts at the conference than Miriam was.

Contrary to popular images of Quakers, Sam was a fighter. And it came as no surprise to those who had worked with him earlier in Quaker and peace groups that Sam fought with every resource available to him from 1972 to 1980 to prevent the passage of unilateral seabed mining legislation that, in his view, would harm UNCLOS III.

Although Sam quickly became known as a fighter on behalf of law of the sea, he also was a reasonable person who earned the respect of many of the people with whom he worked, especially on Capitol Hill and in the administration. Even some of his opponents, including mining industry spokesman Marne Dubs of Kennecott Copper, liked him personally (Dubs 1990). Less dogmatic on law of the sea than he had been as a younger man in regard to world federalism, Sam realized that only a treaty that protected the legitimate interests of U.S. mining firms could ever win Senate approval.

Finally, Sam won high praise from almost all of the people whom Miriam and I interviewed for this book. One congressional staffer, Richard McColl, remembered Sam's "persistence in a positive sense.... Sam always had a large amount of information; he would help us with the arguments" (McColl 1990). Another congressional staffer, David Keeney, offered this assessment of Sam's work in Washington:

> I felt Sam made a very positive contribution to the process. He was very much the diplomat. He realized that he didn't have overwhelming constituent power. He had to work with people and negotiate, and achieve his goals through the power of his arguments.... He was basically arguing on the basis of the rightness of his position, and when you do that, you have to be a diplomat. (Keeney 1989)

State Department officials who worked with Sam on the Public Advisory Committee also praised him highly. John R. Stevenson, a suave New York lawyer who headed the U.S. delegation during the early years of UNCLOS III, recalled that Sam was "one of the best-balanced members" of the PAC. "The hard minerals people were always causing trouble," Stevenson recalled, and Sam helped counter their influence (Stevenson 1990). Another State Department official, Myron Nordquist, remembered that Sam's "personal influence, and the representation of the Quakers and standing for peace ... always had a positive influence, both in our own delegation and as he would wander the corridors and talk to people" (Nordquist 1989). The most detailed tribute to Sam's work on the PAC came from State Department official Bernard Oxman:

> Sam, to my knowledge, never did what I think a lot of people who advocate his position [increased global governance] make the mistake of doing, and that is [saying], for the sake of world order you must sacrifice your other interests. He didn't make people feel that he was

out to make people sacrifice economic interests or environmental interests or security interests for the sake of order. Quite the contrary: he was constantly demonstrating to people how orderly international relationships . . . could help them further their long-range interests, even if there were some short-term problems. Sam also did not make the mistake of assuming that everything the organized representatives of the developing countries demanded (1) was necessarily wise or (2) necessarily had to be treated as a bona fide demand. He had the capacity to see situations as they existed. (Oxman 1990)

The daughter of a Methodist minister who served congregations in the Pittsburgh area, Miriam grew up imbued with the church's social gospel teachings that human society could be improved with divine assistance and that a person's highest calling was to help to improve it. She met Sam when he introduced himself after a talk she had given at a public-speaking contest during her freshman year at Cornell. Her topic, the farmer and world peace, interested Sam, as did Miriam's idealism and her outgoing personality. The twenty-two-year-old graduate student and the seventeen-year-old freshman soon began dating; three years later, in June 1934, they married.

Like Sam, Miriam spent much of her time during the years between 1934 and 1972 working for peace and becoming a leader in Quaker organizations. She was active in many of the same groups, including Friends United Meeting, the United World Federalists, and the American Freedom Association. About 1960, during one of the many AFA-sponsored trips to New York in which she explained the workings of the UN to a busload of winners of high school speaking contests and their teachers, she saw the need for an ecumenical church center at the UN and then worked with Methodists to see the idea through to fruition. Located across the street from the UN, this building housed many of the seminars the Neptune Group sponsored when UNCLOS III met in New York.

Partly because she devoted much time and energy to raising the Levering children at the orchard while also helping raise three motherless children from eastern Europe displaced by World War II, Miriam was less prominent than Sam in peace and Quaker circles from the 1940s through the 1960s. But that situation changed rapidly beginning in 1972, when Miriam had time to showcase her many strengths on behalf of law of the sea.

Many of Miriam's strengths paralleled and complemented Sam's. Believing that much of Sam's and her own previous peace work had not been effective in bringing about concrete changes in international relations, she was determined to do all that she could to help UNCLOS III succeed; she took middle-of-the-road positions on most contentious issues and typically sought compromise and consensus; she was an informed, inspiring speaker; she personified the Quaker ideal of seeking to make the world better; she worked comfortably with people whose primary motivation on law of the sea was academic or secular and with those whose primary motivation was religious; and she spent very little money on herself.

The Leverings' frugality was legendary: while working in Washington, for example, they often slept on the floor of the office in sleeping bags, ate bread and fruit that they had brought from the orchard, and traveled in a dilapidated 1964 Dodge Dart that they had bought used and that accumulated more than six-hundred thousand miles by the end of the 1970s. On one occasion the Dart's right front seat, held in place by a small board and occupied by Congressman Gilbert Gude of Maryland, suddenly collapsed backward as Sam was driving him to a meeting on ocean law in the Washington suburbs. Fortunately, Congressman Gude was not hurt, and the two men remained friends and admirers.

Miriam's frugality, combined with her frequent references to Quakers' international humanitarian work in discussions with delegates, helped convince diplomats and UN officials at the Geneva session in the spring of 1975 that the U.S. Central Intelligence Agency was not funding the Neptune Group, as some suspected it was. Lee Kimball overheard this comment by UN official Gwenda Ward: "Miriam Levering, who won't pay forty centimes to take the bus in Geneva—clearly this person does not work for the CIA." Kimball recalled that "people would think I worked for the CIA, [but] no one ever believed Miriam worked for the CIA" (Kimball 1989).

Miriam had three strengths that distinguished her from Sam and that made her overall contribution even more impressive in retrospect than his was. First, she was an inspiring motivator/friend and hence an effective builder of networks of people who were willing to advance what she and they defined as a common cause. "Miriam was the matriarch of the movement; she mothered us all," Joyce Hamlin of the UMLSP recalled. "She was the calm in the center of the storm, yet also the energizer" (Hamlin 1989).

Jim Orr, who worked on *Neptune* at Geneva in 1975, noted that Miriam's "enthusiasm and commitment are crystal clear to everybody, and I think it's infectious" (Orr 1989). An avid, skillful conversationalist, she took a personal interest in people and made lasting friendships more easily than Sam—or, for that matter, more easily than anyone else I have ever known. Barton Lewis, a founder of the Neptune Group, summed up this strength most succinctly: "Miriam had an amazing ability to work with people of all ilk" (Lewis 1990).

Because of Miriam's charismatic personality and networking skills, the national organization of United Methodist women plus that denomination's UN office enthusiastically joined the crusade for ocean law, and dozens of young people, often working for little or no pay, devoted their summers or took time off from college or graduate school to assist Miriam in her work. And partly because of her personality and ability to make friends—and the outgoing personalities of Kimball, Weaver, Arthur Paterson, and others—the Neptune Group quickly became most delegates' favorite NGO working at the conference.

Miriam's second distinct strength was that she was a prolific, engaging writer. In the U.S. Committee's newsletter, *Sea Breezes,* and in many other publications, she worked hard to make the stories and editorials entertaining as well as instructive. She brought to life the major players on law of the sea in Washington and at the conference, she helped readers identify with the Neptune Group's challenges in making an impact on the issue, and she specified what the reader could do that day (for example, write a U.S. official or member of Congress) to contribute to the cause. Moreover, she often included handwritten notes on copies of publications she mailed out, thus personalizing the link between OEP and its supporters. Many of the scores of letters she wrote each month, largely to supporters and to other NGO representatives, were written with verve and individuality.

Miriam's third strength that Sam did not have was that she was an excellent fund-raiser, especially among individuals who read *Sea Breezes* and other publications or ones with whom she had established a personal connection in other ways. Besides larger contributions from wealthy individuals and foundations, Miriam raised perhaps $10,000 to $15,000 each year from 1973 through 1981 from individuals and from local churches (prompted by individuals within them) who gave to OEP because they trusted that Miriam and Sam would put the money to good use. Miriam regularly sent personal thank-you notes to acknowledge each gift and, in many cases, to state precisely how it would be used.

In short, the Leverings brought to their involvement in law of the sea many years' experience in peace and church work, an intense desire to help UNCLOS III succeed in building international law and institutions, and a combination of strengths (for example, Sam's expertise and Miriam's interpersonal skills) that made their combined contribution far greater than either alone could have achieved. In this case, at least, NGOs were indeed the lengthened shadows of those who gave them life and direction—including Sam's and Miriam's many able associates as well as themselves.

This Study and the Existing Scholarship

Third and last, where does this book fit into the scholarship on NGO involvement in international negotiations? As a historian of U.S. foreign relations accustomed to a vast literature on such topics as the origins of the cold war and the Vietnam War, I was surprised to find only a small body of writing on NGOs in international negotiations, and a not much larger one on the broader topic of NGOs in global politics generally. Indeed, I found only four books in English—all edited collections of articles, not detailed studies—that focus on NGO and INGO involvement in contemporary world politics.[8] Thus this book, to my knowledge the first detailed case study of an NGO's involvement in a UN conference and in an important capital, is breaking relatively new ground.

More broadly, however, this book fits into a large body of scholarship in international relations, elaborated during the 1970s and 1980s, that argues that the "realist" theory of world politics, which contends that powerful nation-states are the only significant actors on the global stage, is grossly inadequate in understanding a world characterized by what political scientists Robert O. Keohane and Joseph S. Nye call "complex interdependence." The seminal early statement of the "interdependence" theory is contained in the authors' 1977 book *Power and Interdependence: World Politics in Transition,* which also contains a perceptive discussion of the "oceans issue area" as an example of this interdependence. In a later edition, Keohane and Nye

---

8. Listed by date of publication, these are Willetts 1982, Princen and Finger 1994, Weiss and Gordenker 1996, and Smith, Chatfield, and Pagnucco 1997. Besides the few books, more articles and working papers have appeared on NGOs active in global politics. Two valuable examples are Barnes 1984 and Smith, Pagnucco, and Romeril 1993. The latter provides an insightful overview of the field and contains an excellent bibliography.

describe "complex interdependence" as "a situation among a number of countries in which multiple channels of contact connect societies (that is, states do not monopolize those contacts); there is no hierarchy of issues; and military force is not used by governments toward one another" (Keohane and Nye 1989, 249).

These three conditions existed during UNCLOS III, though the threat of force remained as an implicit alternative if agreement was not reached on such key issues as rights of passage through straits and archipelagoes. Also existing—and evolving—before, during, and after the conference was what Keohane and Nye call a "regime" of formal rules, informal restraints, and widely accepted practices that helped limit international conflict over the oceans (1989, 92–96).

This study also connects (a) with a well-established literature on movements for social and political change within nations ("social movement theory") and (b) with more recent efforts to develop a theory of transnational social movements operating within the global political system.[9] Scholars who work in these closely related areas have noted that a variety of "political opportunity structures" within the different environments in which social movements work either "facilitate or constrain social change efforts."[10] In this case study, the members of the Neptune Group faced very different "opportunity structures" in Washington and at the conference that changed over time and that influenced the effectiveness of their efforts.

In a 1993 monograph, Jackie Smith, Ron Pagnucco, and Winnie Romeril defined "transnational social movement organizations" (TSMOs) as "those NGOs that target international institutions and attempt to affect international policies in their ultimate aims of influencing state behavior. . . . TSMOs also differ from national SMOs in that they have memberships that are not contained within the boundaries of a single state. This provides them with political resources that may not be available for national SMOs" (1993, 2–3).

---

9. A useful introduction to social movements in national settings is Tilly 1984. On transnational social movement organizations, see Willetts 1984, 1–27; Smith, Pagnucco, and Romeril 1993, 1–19; and especially Smith, Chatfield, and Pagnucco 1997.

10. The quotations are from page 2 of a draft outline for the introductory chapter for a proposed book on transnational social movements and world politics, sent by Charles Chatfield to the author in July 1995. This book was eventually published as Smith, Chatfield, and Pagnucco 1997.

The Neptune Group should be considered both an SMO and a TSMO. Its organizations were national in that (a) virtually all of their personnel and resources were American and (b) they sought primarily to affect U.S. policy—not the policies of other governments—through their work in Washington and through the mobilizing of supporters within the United States. But the group also was transnational because the Leverings used their well-placed, respected Quaker connections in New York (Quaker House and the Quaker UN Program), in London (the Friends World Committee for Consultation), and in Geneva (Quaker House) to give their work at UNCLOS III added credibility and resources.[11] Lee Kimball also established important connections with non-U.S. NGOs and research institutes, especially in western Europe and Latin America.

Although such concepts as "complex interdependence," "political opportunity structures," "international regimes," and "transnational social movement organizations" help clarify the contours of contemporary world politics, it must be remembered that most NGO representatives who enter the national and global political arenas make an impact—or fail to do so—as much because of their personal characteristics, their approaches to their work, and the specific, changing circumstances affecting their efforts as because of institutional affiliations. This generalization might not be as valid for such contemporary TSMOs as Greenpeace, whose reputation for innovative, controversial tactics stands higher than that of any person within it. But it certainly held true at UNCLOS III not only for the newly formed organizations that made up the Neptune Group, but also for more established NGOs like the Sierra Club International and the United World Federalists.

At the conference and in Washington, what was valued among NGO representatives was what individuals and small groups of people working together could contribute, and not the organizational badges they wore. As UNCLOS III thus shows, contemporary social scientists and historians need to reserve a large place in their thinking for the knowledgeable, shrewd, and determined individuals, both nongovernmental and governmental, who have influenced the course of history in the past and who continue to do so today.

11. Besides these and other Quaker organizations involved in transnational affairs, individual Quakers played leading roles in establishing such TSMOs as Greenpeace, Oxfam, the International Fellowship of Reconciliation, and Amnesty International. See Smith, Pagnucco, and Romeril 1993, 6–7.

2

Getting Started and Mobilizing Influence

THE STORY OF THE NEPTUNE GROUP began in January 1972 when dormant apple buds on the Levering family's commercial fruit orchard on a Blue Ridge mountainside in southwestern Virginia were frozen and brown. The temperature had fallen in two nights from forty-seven above to seven below. Cold-damaged apple buds were sadly familiar in April, but never in January. Experts predicted little or no crop.

Then the phone rang. Edward F. Snyder, executive secretary of the leading Quaker lobbying group in Washington, the Friends Committee on National Legislation, had a question for my husband, Samuel R. Levering, who headed the FCNL's Executive Committee. "Should the FCNL," he asked, "accept the request of two Philadelphia Quaker businessmen, A. Barton Lewis and William F. Fischer Jr.?" The Philadelphians had offered to fund a one-year special project to help build support for a comprehensive, equitable law of the sea treaty. Sam responded that he not only approved the idea, but also that, because of the freeze on the orchard, he would be available to work on it if Lewis and Fischer wanted to hire him—so long as he could include me as his unpaid coworker.

Like Sam and me, Lewis and Fischer had long been committed to strengthening the rule of law worldwide, and both had read the 1970 U.S. draft convention on law of the sea that the Nixon administration had proposed to the forthcoming UN Law of the Sea Conference. They immediately recognized its potential for world order and equity, as well as its vast complexity and the difficult, divisive questions it raised. Lewis read the draft treaty first and, to hear Fischer tell it,

Lewis phoned excitedly and gave him fifty-five minutes to read it and meet him for lunch to plan their next move!

The U.S. draft treaty proposed sharing revenue from oil royalties beyond the two-hundred-meter offshore depth with the international community, and especially with developing nations. Not surprisingly, oil companies were cool to the idea, as were mining firms interested in harvesting the hard minerals of the deep seabed. They also disliked the proposed management system, an "international seabed authority" set up to serve the broader interests of humanity. Lewis and Fischer called this draft "Louis Sohn's treaty" because the eminent Harvard Law School professor and hero to supporters of international institutions had spent a couple months in the summer of 1970 helping write the draft treaty. Other State Department officials—notably under-secretary Elliot Richardson and Bernard Oxman, who chaired the drafting committee—shared Sohn's desire to strengthen global governance. And so did at least one other committee member, Leigh Ratiner of the Defense Department. The Philadelphians saw their support for the draft treaty as a concrete way to enhance the prospects for world order under law.

By late January Sam had been hired to head the one-year effort, and he and I began work in a small FCNL office early the next month. On February 8 Sam, Ed Snyder, and Robert H. Cory, director of a Quaker study center near Capitol Hill (the William Penn House), met with the two Philadelphians. They agreed to hire Sam and me to work—at first under FCNL auspices—to foster the treaty effort. We would seek to bolster a wavering administration, inform Congress of the treaty's potential significance, and prevent unilateral actions in Congress that could preempt in Washington what the administration had offered to negotiate with the world community at the UN. Our ultimate objectives were to prepare the Senate for eventual consent to the treaty, and then to make the treaty system work.

We had barely settled into our office on the third floor of the FCNL building on Capitol Hill before we faced hearings on the Deep Seabed Hard Mineral Resources bills that lawmakers close to the American Mining Congress, an industry lobbying group, had introduced into the Senate and the House of Representatives. These bills would have undermined the U.S. approach to the negotiations and preempted this international effort to fashion an equitable solution. To our surprise, the bills had no organized public opposition.

As detailed in the next chapter, we set out to organize it. Opposing this legislation became our first vehicle for publicizing the prospective treaty.

Sam and I plunged into learning the formidable details of the four treaties produced by the 1958 and 1960 UNCLOS conferences. But it was their failure to settle the limits and content of coastal states' control, combined with the new prospect of deep seabed mineral wealth, that necessitated the third conference (1973–82).

We studied the preparatory work of the UN Seabed Committee since 1968. We sensed that ahead lay painful tradeoffs between landlocked and coastal states, between countries that benefited from dryland mining and those containing would-be seabed miners, and between nations with naval and marine-research interests that wanted the right to go close to foreign shores and those coastal nations that were suspicious of "spying." We wondered whether roughly one hundred developing nations could hold together in their UN "Group of 77" bloc when their geography dictated differing policies. With so many diverse interests, could a treaty be agreed on by 1974–75, as conference planners hoped, or indeed ever?

We also began to acquaint ourselves with the congressional maze, the key players both nationally and internationally, and the interests of the major ocean industries involved. We read the legal experts until our Yankee/Appalachian speech acquired a legalese that, for example, transformed "land grabs by coastal states" into "creeping jurisdiction." Observing the fights over U.S. policy on ocean law within the administration and in Congress, we learned that policy debates can be as intense in some capitals as they are in international forums.

Our lives would be occupied for ten years by the high-grade manganese nodules found three miles down on the tropical Pacific floor. (A mining company kindly supplied me with samples, which I often held up or passed around to stir up interest during my speeches.) The nodules attracted multinational corporations and congressional and UN action because they contain nickel, copper, cobalt, and manganese. The burnt-potato-like lump develops over millions of years in a habitat so cold that little can live—so cold, in fact, that one clam species takes two hundred years to reproduce!

Although the nodule is widely scattered on the seabed, the areas of primary commercial interest lie between Hawaii and Mexico. Four U.S. firms—Kennecott Copper, Tenneco, Lockheed, and U.S. Steel—were early pioneers. In the middle 1970s Howard Hughes, the reclusive

Manganese nodules from the ocean floor on the deck of an experimental mining ship, mid-1970s. Courtesy of Deepsea Ventures, Inc.

billionaire then under contract to Lockheed, added his famous dredge, the *Glomar Explorer,* to the search for nodules. One company, Tenneco's Deepsea Ventures, filed a claim in 1974 with the State Department for a mid-Pacific mine site; the claim was rejected on the grounds that the United States lacked jurisdiction.

Without the dispute over nodules, the conference would have been shorter, many of the developing countries and Canada less worried about competition with their own land-based hard minerals, and the nascent seabed mining industry without major involvement in the conference and on Capitol Hill. Nor would we have spent a good portion of a decade analyzing political alternatives and economic implications of seabed mining, organizing witnesses for a seemingly endless series of congressional hearings on the subject, and shakily facing representatives of the influential, well-funded American Mining Congress.

Sam hastily flew to Boston to get advice from Professor Sohn. His practical suggestion was to find allies and work with them. Most of the allies Sohn had worked with in 1970 on the draft treaty were still in the administration working on law of the sea. Sam visited with Bernard Oxman and Stuart McIntyre, among others, in the State Department; Leigh Ratiner, then in the Interior Department, and Vincent McKelvey of the Geological Survey. He also contacted the State Department's Myron Nordquist, who had begun working on the issue after 1970.

After doing some reading and discussing with me what he had learned from the experts he had consulted, Sam concluded that, legally, the oceans were a near-disaster area akin to the American Wild West of the late nineteenth century. Time, new technology, and national decision making had made obsolete the traditional national three-mile limit set by Dutch scholar Hugo Grotius in the seventeenth century. A new law of the sea, with adequate institutions and mechanisms to enforce it, was desperately needed.

As longtime crusaders for global governance, Sam and I saw the treaty-making process as an excellent opportunity to try to strengthen international law and world order over two-thirds of the world's surface. An opportunity of this magnitude was unlikely to happen again soon—certainly not in our lifetimes.

Finding Allies and Resources

During the spring of 1972, while Sam lined up protreaty witnesses for the congressional hearings and spread internationalist memoranda on Capitol Hill, I went door to door among religious, world order, scientific, public interest, environmental, and general foreign-policy groups. We also organized a small, secular, independent lobbying organization, the United States Committee for the Oceans, in early 1972. Sam served as executive secretary (he and I thus doing much of the organization's work), while former United States Supreme Court justice Arthur J. Goldberg and former governor Russell Peterson of Delaware became its honorary chairmen. Aided by college and graduate students, Sam launched a campaign to educate Congress on what we saw as the important U.S. stake in updated, truly international ocean law.

We soon discovered that not all the roadblocks to progress were in Washington. The slow-moving preliminary negotiations needed help. Public education, including media work, was essential to advance our cause. In January 1973 we presented the need for public education to the board of directors of the American Freedom Association, a North Carolina-based world-order group in which the two of us had long been active. When the board agreed to sponsor an Ocean Education Project, with me as executive secretary, our educational program got under way. There was another reason for starting OEP: the U.S. Committee, as a lobbying group, could not receive tax-deductible contributions. Having OEP enabled us to receive the tax-deductible contributions (from foundations as well as from individuals) that made possible our ongoing educational efforts and work at UNCLOS III. From 1973 forward, all our funding except the small amounts needed to cover the U.S. Committee's work with Congress came to OEP.

We confronted the kinds of problems that most fledgling NGOs face. With limited resources, how and where should we concentrate our efforts? Who was the audience for our publications? Where could we find funding? Gradually, many of the answers came.

Robert Cory, a Quaker active in many causes in Washington, let us use the William Penn House for monthly luncheons with congressional staff members (normally our allies on the Hill), media people, and other NGO representatives who supported our work. Periodically, Quaker House in New York served a similar function. Given Sam's and my interest in the ocean environment and in stronger international institutions, OEP made common cause with the international programs office of the Sierra Club's Office of International Environmental Affairs, led in the early 1970s by Patricia Scharlin-Rambach, and with the oceans committee of the United Nations Association of the United States of America (UNA-USA), chaired by Lili Hahn and Eleanor Schnurr, UN representatives respectively for Unitarian-Universalist Women and for the American Baptist Church. Sam and I wrote for numerous publications, gave endless briefings, and raised countless questions in meetings on ocean law. We sought to get law of the sea onto the programs of such national conventions as the League of Women Voters and the National Science Teachers

Association, and we frequently were successful. I offered information to teachers in the journal of the National Education Association, and hundreds of requests poured in and were answered.

Our most serious and persistent problems were financial. We never succeeded in raising the substantial sums that a large-scale public-relations and educational campaign would have required. A few foundations (notably the Lilly Foundation and the Rockefeller Foundation) gave us modest grants; hundreds of people made contributions that usually ranged from $5 to $500. And we depended heavily on the work of volunteers, on joint ventures with more affluent NGOs (for example, the Sierra Club) that had facilities we needed or money to trade for our specialized experience.

We also received help from unexpected sources. A couple years into our work, a young woman who had been a volunteer asked Sam and me to dinner in her Washington apartment. After dinner she quizzed us at length about our goals and plans, and then said: "My brother committed suicide. He left me $14,000. I know that he would want you to have it."

Another example of what I call the pretty hand of Providence occurred at the last session of the UN preparatory conference in Geneva in the summer of 1973. I arrived there inexperienced and without staff, my college French having mostly disappeared. In Geneva I stayed at an inexpensive accommodation used mainly by international students, the Foyer John Knox. En route to the shower one day I met Jim Bridgman, a young American who needed to gather information for a paper for his Beloit College international-relations seminar. I needed an aide, especially one fluent in French, as he was. He wrote his paper after we jointly visited every national UN mission in Geneva, sharing our viewpoint with them. Thus began an eight-year-long relationship at many of the conference sessions with Jim contributing his time, his expenses, and his burgeoning knowledge of the issue.

A college student who made an even greater contribution to the Neptune Group was Arthur Paterson, a multitalented young man with an exceptional personality that helped him develop valuable friendships both at the conference and in Washington. A senior at Wesleyan University when he first contacted us early in 1974, Paterson worked off and on for OEP and UMLSP for the next eight years. Extremely dedicated, he often put in very long hours for little pay. Arthur excelled not only in his work at the conference, but also as an organizer of meetings and as a writer and editor for several Neptune Group publications, including *Neptune* and the main UMLSP/OEP publication, *Soundings*.

During the 1975 session in Geneva, I invited a woman whose brother I knew to tea. As she said good-bye, she remarked, "I remodeled the carriage house in Voltaire's chateau in Ferney [near Geneva]. I would like to have you stay in it with me." She thus provided free lodging for me during the remainder of the 1975 conference, and also during subsequent sessions in Geneva.

In Washington, Sister Mary Beth Riessen, a Catholic high school teacher who had been attending our William Penn House luncheons, offered to join our staff for the subsistence required by the order: $300 a month. Through her speaking, writing, and personal calls, she became the portal to the Vatican's delegation to the conference and to many U.S. Catholic groups.

But nothing could compare with the long-term benefits that resulted from my walking into one blue-carpeted office near the capitol. I told my law-of-the-sea story to Joyce Hamlin, the soft-spoken director of the United Methodist Women's legislative office, and asked, "What can you do?" She poured me some coffee and listened graciously. In an interview many years later, Hamlin recalled the impact of my visit: "Dozens of people came into my office [asking for support], but this one had a special ring to it. Who ever thought of the nodules on the seabed floor? It was exciting to think we might be able to keep [the seabed floor] from being carved up" (Hamlin 1989).

Sometime later Joyce walked into our office, looked around, and asked, "Do you need an extra phone here? Would you like a typist? Some research assistance?" She handed us a check for $2,000 and asked, "Could you come on Thursday afternoon to talk about working together at the conference?" When we went, we learned that a woman had recently given $100,000 to the Women's Division. Joyce and Mia Adjali, codirector of the Methodist UN office in New York, had obtained $25,000 from the gift to launch work on law of the sea.

This United Methodist grant, followed by sustained funding thereafter, made possible the publication of a professionally printed newsletter, *Soundings*, that went out to a large mailing list composed primarily of Americans—officials, journalists, supporters of Sam's and my organizations and the Methodist project, and others. It also generated the publication of *Neptune*, our independent newspaper for conference delegates, and the coalescence of the Neptune Group of

NGOs (the U.S. Committee, OEP, and the UMLSP) in Washington and at the conference. The Methodist connection gave us access to this large denomination's national staff, its media specialists, its publications, and a significant number of activists and academics. And because the Methodists owned the church center across from the UN, the connection with the Methodists gave us offices convenient to the conference (most of the sessions were held in New York) and space in the church center for a steady succession of seminars, luncheons, and receptions for delegates, journalists, and other NGOs.

No other church, nor in fact any other public interest or foreign-policy NGO, came close to the United Methodists in the intensity of focus and breadth of outreach on ocean law and stewardship. None committed comparable resources to this effort. Their informational services ranged from the most sophisticated analyses of the negotiations and the U.S. legislative process—written by Lee Kimball and Arthur Paterson, among others—to explaining the issue to local newspaper editors, and even down to teaching kindergarten children how to carve a whale out of soap. Without the resources and personnel the Methodists contributed, there would have been no Neptune Group.

The Group's Two Pillars

Sam and I, plus Sister Mary Beth Riessen and numerous interns and volunteers, formed one pillar of the Neptune Group. The other pillar was the UMLSP. Barbara Weaver and Lee Kimball—the attractive, engaging leaders of the Methodist project at the height of its work at UNCLOS III—were a full generation younger than Sam and I were, and thus were contemporaries of many of the "twenty-something" delegates to the conference. Both were deeply committed to the achievement of a treaty that would reflect Third World as well as Western interests and concerns. Otherwise, however, Lee and Barbara were quite different, bringing complementary skills to the Neptune Group's work.

A native of Washington state with an undergraduate degree from Stanford, Lee was highly intellectual in approach and had, she recalled, a "secular, at times cynical style" that helped her establish effective relationships with some of the more hard-nosed delegates and mining-company representatives (Kimball 1995). Especially in the early years, she was embarrassed to admit to inquirers that she worked for a church-related organization. Lee had done graduate work in international relations and law of the sea at the Johns Hopkins School of Advanced International Studies in Washington during the two years before her involvement with the group, and had developed a conceptual, analytic approach to the issues surrounding the conference.

In contrast, religious yet fun-loving Barbara, a native midwesterner, came to law of the sea from public-school teaching and Methodist Church work in Columbus, Ohio. Her approach emphasized Christian stewardship and equitable sharing of the world's resources, and focused on a challenging question: how could ordinary people—especially church people—be encouraged to become sufficiently interested in the conference to take an effective stand on issues involved in the negotiations? At the peak of her involvement from 1976 through 1979, Barbara, an excellent speaker who never bored an audience, used visual aids, issues of *Soundings,* regional seminars, "sea suppers," devotional materials, and local Methodist activists to stimulate and involve relatively many church people, primarily Methodists.

Within a year after becoming involved with Sam and me in 1974, Lee had joined UMLSP (which had the funds to pay her a salary) and had blossomed into one of the Neptune Group's stars. Even more than Sam, the serious, "businesslike" Lee developed a detailed knowledge of the substantive issues in the negotiations, especially those relating to seabed mining. From me she learned to shed some of the academic prose that hindered effective dissemination of her ideas. By the time the conference held a session in Geneva beginning in March 1975, she was the editor of *Neptune,* a role that would continue throughout the conference. She also learned that, while factual knowledge and seriousness of purpose were important, it was equally necessary to establish personal relationships with those whom one might wish to influence in the future. Lee emphasized what she had learned from me in a 1980 letter:

> It is tough to forget my first meeting with [you], which took place at an evening sponsored by the Carnegie Endowment's Project Dialogue in 1973. You came up to me afterward and said something about how much I knew about the Law of the Sea, "and so pretty too!" That was my first exposure to your highly effective technique which I later learned to adapt under your tutelage—leading with a compliment, rendering the personal touch and thus laying the groundwork for sustained future contact. (Kimball 1980a)

Lee made valuable contributions in all areas of the Neptune Group's work. She worked closely with Sam in lobbying Congress and consulting with the administration; she labored with Miriam and others on publications and traveled with Barbara to speak in Methodist forums across the country to build public support for the treaty; and she developed into the most effective NGO representative at the conference.

Lee clearly made her greatest contribution at the conference, as both delegates and NGO colleagues repeatedly commented in interviews for this book. State Department official Otho Eskin found her to be "very articulate and effective" (Eskin 1989); Edward Miles, a professor at the University of Washington who attended many sessions of the conference, noted her "extraordinary access" to delegates (Miles 1990); and Arthur Paterson recalled her skill in "helping to identify a problem and bringing in outside experts" (Paterson 1989). Tommy T. B. Koh, Singapore's chief negotiator who became president of the conference, described Lee as "quietly competent," "a solid pro," "very hard working," and "very tactful and discreet" (Koh 1990).

Eleanor Smith, a Neptune Group intern who worked with Lee at conference sessions in 1980 and 1981, remembered her as "an extremely quick thinker, able to cut through the muddle to the heart of an issue.... She's also savvy. She knows the pulse of a situation, and who should be contacted to get a particular thing done" (Smith 1989). Smith elaborated on her contribution to the conference:

> Lee was respected at the conference as being one of the few who really understood the implications of the treaty process, the policy issues that had to be decided, and the technical aspects. And that is no small feat, given the nature of this conference. She was regularly sought out by most delegates. The impression was that she was the pivotal person to go to see if you wanted an answer. If you were a delegate, you had to save face; you couldn't go and ask another delegate what something meant, because then it would appear that you didn't know what you were doing. But you could easily go to see Lee Kimball, who represented a non-governmental interest and who was of the greatest integrity, and you knew that she wasn't going to turn around and tell the next delegate that you came to her to find out what something meant. So she was a walking encyclopedia of information for delegates at the conference. (Smith 1989)

Inspiring, practical Barbara Weaver and discreet, analytic Lee Kimball thus were almost ideal complements in UMLSP. Besides ensuring that national Methodist organizations funded her project, Barbara put together learning materials and set up programs in local churches and in Methodist district meetings across the country. Although Lee traveled with Barbara to Methodist meetings, she devoted most of her time to issues facing the conference. Arthur Paterson remembered Barbara as "a charmer of the diplomats, an educator and charmer of the people [Methodists] that she worked with, a very good network organizer" (Paterson 1989). Together and separately, Barbara and Lee contributed substantially to the Neptune Group's effectiveness.

The Larger NGO Effort

The Neptune Group operated as part of a larger NGO effort at UNCLOS III, composed mostly of Americans but also including a few activists from western Europe and a few from other parts of the world. This effort was not as large or as well publicized as were NGO activities at the UN-sponsored conferences at Stockholm in 1972 on the environment and at Rome in 1974 on food.[1] But neither was NGO involvement as controversial as it had sometimes been. Especially after an initial flurry of NGO publicity seeking at Caracas in 1974, NGO representatives generally did not anger delegates and UN officials—as they did especially at Stockholm—by holding press briefings and alternative meetings to denounce policy positions that failed to meet their demands.

Partly because of Stockholm and partly because of the behavior of some NGO representatives during the sessions in Caracas in 1974 and at Geneva in 1975, many UN officials and delegates disliked NGO representatives as a whole and wanted to limit their participation in the conference. Accordingly, NGO representatives were not invited to address plenary meetings after the Caracas session, and they were excluded from negotiating meetings after the Geneva session (Miles 1990). The coolness of many UN officials and delegates, combined with the length and complexity of the negotiations, made UNCLOS III inhospitable to NGO representatives and journalists who were not willing to undertake the long-term effort needed to become accepted and knowledgeable at the conference.

1. On the Stockholm conference, see Todd 1973, 42-45, and Feraru 1974, 31–60. On the Rome conference, see Weiss and Jordan 1975, 268–71.

Although fifty-five international NGOs were represented officially at one or more sessions of the conference, in fact the number of international or national NGOs whose representatives undertook an ongoing, active role in UNCLOS III was much smaller, perhaps less than ten. The two main interests of the activists were world order (including law/governance and redistribution of wealth from rich to poor nations) and the maritime environment. As the conference dragged on into the late 1970s and early 1980s, activists associated with the Neptune Group—including a few Europeans—probably made up more than half of the public-interest NGO representatives working at the conference at any given time.[2]

Americans were especially prominent in the NGO effort for several reasons. First, environmental groups and organizations supporting the development of international institutions generally were better organized and funded in America than in other countries in the 1970s. Second, the openness of the U.S. government to public-interest group involvement on law of the sea stood in marked contrast to the situation even in some other democratic countries, notably Great Britain, where officials openly discouraged NGO involvement. Third, in the American political system NGOs and economic interests could play the administration off against the Congress, and vice versa, more effectively than in parliamentary systems.[3] Danish delegate Peter Bruckner recalled that in western Europe in the 1970s "there was no NGO with the same sort of broad-range interests as the Neptune Group" (Bruckner 1990).

Environmental groups generally were more successful in achieving their goals than were world-order organizations, with the major exception of the Neptune Group. During the early years of the negotiations, the contributions of environmentalists, both in Washington and at the conference, were modest. But in the late 1970s,

---

2. Lacey 1982, 100-5. Although most of the information in this paragraph is drawn from Lacey, the estimate of ten active NGOs is ours. Estimating the number of active NGOs is difficult because many NGO representatives were accredited by international NGOs that themselves had little or no interest in the conference. Lee Kimball, for example, was accredited by the Women's International League for Peace and Freedom, whose leaders and members generally took very little interest in UNCLOS III.

3. A brief but perceptive discussion of NGO involvement is contained in Schmidt 1989, 63-67. See also Sanger 1987, 32–33.

with the encouragement of U.S. officials such as Elliot Richardson and members of the Neptune Group, several capable environmentalists—including Jim Barnes of the Center for Law and Social Policy, Patricia Forkan of the Humane Society, and Anita Yurchyshyn of the Sierra Club—became members of both the Public Advisory Committee and the U.S. delegation to the conference and helped (a) set U.S. policy and (b) draft the treaty's environmental articles relating to marine mammals, seabed mining, and other subjects. During this effort, discussed in greater detail in chapter 5, these and a few other environmentalists offered a model of how to be effective in international negotiations: know the subject and the negotiating history thoroughly; work quietly but persistently within the system; and be willing to accept achievable improvements instead of holding out for the ideal.[4]

The NGOs that tended to be ineffective during the negotiations were largely the world-order, pro-developing-nation NGOs with ambitious visions of how the conference should ensure that the new law of the sea gave the main economic benefits from the ocean, especially from oil and mining, to the poorer nations. These benefits, they argued, should come not only from the area beyond the two-hundred-mile exclusive zone—a position that was widely supported—but also between the twelve-mile territorial zone and two hundred miles, where large, easily recoverable reserves of oil and natural gas were concentrated. Among those who advanced this broad conception of global revenue sharing—often long after it was clear to everyone else involved in the conference that such ideas were not favored by a single coastal state and hence had no chance of being accepted—were Americans Richard Hudson of the War/Peace Institute (New York) and John Logue of the World Order Research Institute at Villanova University (near Philadelphia).

When Sam and I were getting started in 1972 and 1973, we were strongly attracted to the idea of obtaining revenue from the "common heritage" for the world's poorest nations, and Sam invited Logue to Washington several times to testify against unilateral seabed mining legislation. During the session in Caracas in the summer of 1974, however, we realized that Logue's ideas on revenue sharing had already lost out. We distanced ourselves from him when we noticed that he was a poor listener and that delegates at the conference

    4. Richardson 1990, Forkan 1989, Curtis 1990, and Yurchyshyn 1990. For more details on Yurchyshyn's contribution, see Irwin 1982, 142-43.

avoided him. "John Logue became a joke," U.S. negotiator Leigh Ratiner recalled. "You couldn't any longer listen to John Logue because he was monolithic in his left-wing view" (Ratiner 1989). Delegate Tommy Koh of Singapore noted that Logue had "grandiose, idealistic ideas" that were "politically not possible to sell" (Koh 1990). By the mid-1970s Sam and I were embarrassed that Logue was accredited to UNCLOS III by a movement, world federalism, in which we had long been active and in which we still believed.

Elisabeth Mann Borgese, the highly intellectual daughter of novelist Thomas Mann who headed a respected international NGO, Pacem in Maribus, likewise alienated many delegates by lecturing them on precisely what provisions the completed treaty should contain, including generous transfers to poorer nations of revenues from oil and other minerals. She also amused and spurred on members of the Neptune Group at early conference sessions by telling us that, compared with her, we were amateurs.[5] Although partly true at the time, such comments—as well as her long, prescriptive speech at a plenary meeting in Caracas—were tactless. Having lost much of her effectiveness as an NGO representative by 1975, she sought to influence the negotiations by becoming a delegate from Austria and by asking to publish articles in our publication, *Neptune*.

The Neptune Group soon realized that the efforts of Logue, Hudson, Borgese, and others to persuade the conference of the need to redistribute wealth from the ocean were doomed to failure. We saw these overly idealistic internationalists slipping into irrelevance—a fate we hoped to avoid by seeking to assist the delegates in their difficult negotiations instead of lecturing them from a presumed moral or intellectual higher ground. We still believed in what Malta's UN Ambassador, Arvid Pardo, in an electrifying 1967 speech at the UN had called "the common heritage of mankind"; but we recognized by late 1974 that any financial benefits for the poorer nations would have to come from the international area beyond two hundred miles (Sam Levering 1990).

5. Levering 1990, Koh 1990; Ratiner 1989; and Paterson 1989. During our interview with Ratiner, Miriam Levering commented about Elisabeth Mann Borgese: "We wanted to separate ourselves miles from her, because we felt she was looking at things from an abstract, academic, ivory-tower point of view up there in Dalhousie [Nova Scotia]. We were impacted by the fact that we sat right across from the U.S. Senate."

Although we and other Neptune Group members abandoned some of our early hopes for the conference, more than enough of the original vision remained to sustain our work. Even if imperfect from our standpoint, a comprehensive law of the sea, complete with new international institutions and new procedures for settling disputes, could be a giant step toward global governance. In contrast to the largely negative effort to stop the Vietnam War that had been the American peace movement's major focus in the late 1960s and early 1970s, work on ocean law represented the chance of a lifetime for those involved to try to make a positive contribution to world peace that might well last for generations. What we needed to do in Washington to bolster the treaty process quickly became clear; our role at the conference emerged more gradually, but with equally gratifying results.

3

Opposing Unilateralism in Washington

THE EARLY AND MIDDLE 1970s were exciting times for Sam and me and our associates to be working in Washington for ocean law. Many other people in the capital—members of Congress, administration officials, other NGOs, and media people—shared our belief in international cooperation and the rule of law. Thus, although we always had opponents, we also had numerous allies who encouraged and supported us in our work. Most significantly, with modest but persistent help from us and from like-minded NGOs, our allies in the administration and in Congress ensured that unilateral seabed mining legislation—legislation that might well have destroyed the conference if it had passed early in the negotiations—was not enacted until 1980, by which time the conference's major compromises on deep seabed mining and other issues had already been achieved.

An example of the support Sam and I found so heartening occurred in October 1974. Early that month Lee Kimball, who was then working full-time for us, contacted editorial writers at the *Washington Post* and brought them up to date on developments at the conference and unilateralist viewpoints in Washington that threatened the negotiations. The resulting lead editorial on October 15 sounded very much like the analyses we were providing in our own publications at the time:

> [T]he hardening of the [200-mile] zone concept [at the conference] makes it all the more important to ensure that the resources of the deep seabed "common heritage" are fairly exploited and that all of the other issues of the seas—fisheries management, environmental

controls, navigation, research and the like—be worked out on the basis of agreed international rules. The role of congressional restraint in making this possible is absolutely critical.[1]

This editorial extolling international consensus seeking thrilled us, and we also were pleased when the *Post* published a letter from Sam praising the editorial on October 26 and when *Post* reporter George C. Wilson frequently talked with us when preparing articles on ocean issues. Wilson first put our concerns on the front page of the *Post* with an article on May 6, 1973, headlined "Battle Stirs over Seabed Mines Bill" (Wilson 1973, 1A). Although we were Davids compared to the Goliaths of lobbying groups in Washington (for example, the Petroleum Institute and the AFL-CIO), we discovered repeatedly that we were making a contribution on law of the sea.

Fighting Seabed Mining Legislation

There were two main reasons why some members of the Neptune Group trekked around Capitol Hill carrying memoranda, legislative analyses, drafts of resolutions and amendments and testified and developed strategies for a decade (1972–82). First, several positive goals inspired us: we were determined to gain support for a treaty that would be fair to both developing nations and maritime powers; that would decrease pollution, protect fish, whales, and other ocean life; that would protect responsible freedom of navigation; that would reduce conflict by settling the issue of ownership of ocean resources; that would, in short, establish effective law for the oceans and institutions for hard-mineral management and dispute settlement. Second, one negative goal spurred us: we were determined to prevent any unilateral U.S. action that would damage the prospects of obtaining an equitable treaty.

During the 1970s Congress faced three sets of unilateral legislative initiatives that challenged these goals. The first was an attempt by the

---

[1]. "Congress and the Law of the Sea," *Washington Post*, 15 October 1974, 20A. Whereas the *Post* agreed with our basic position and printed Sam's letter to the editor, the *Wall Street Journal* strongly disagreed with our internationalist viewpoint. See, for example, "Enough's Enough," *Wall Street Journal*, 17 December 1973, 14A. Unlike the *Post*, the *Journal* did not print a letter Sam wrote in response to the editorial.

corporations that belonged to the American Mining Congress to possess the best sites for deep seabed mining before the LOS treaty would enter into force. The second was an attempt by the U.S. coastal fishing industry to establish control over fishing out to two hundred miles. The third was an effort to establish U.S. control over ocean pollution within a similar two-hundred-mile zone.

All of these goals were to be achieved by unilateral U.S. actions, disregarding previous U.S. commitments in treaties or declarations. For eight years, Sam's U.S. Committee and his close allies inside and outside the Neptune Group opposed these attempts to undermine the treaty process.

The brainchild of the American Mining Congress, the Deep Seabed Hard Minerals Resources bill, first introduced in Congress in 1971, was designed to license U.S. companies to mine blocks of the ocean bottom beyond two hundred miles before the treaty came into force. The Nixon administration opposed this bill because it wanted a new, comprehensive law of the sea and did not want to undermine the negotiations. U.S. officials found it hard to ease the international community's fear that the bill's "interim legislation" might become permanent, that an alternative mining system would be set up through the "reciprocating state" provisions in the bill under which several mining countries, by recognizing each other's claims, would bypass the UN treaty process and take over the international seabed.

Given mounting pressure to pass a deep seabed mining bill, protreaty officials in the administration must have longed for some powerful lobby to counter the American Mining Congress and its allies—the National Association of Manufacturers flanked by the AFL-CIO, for example! Instead, an orchardist walked into the State Department, speaking for a small religious and internationalist constituency. Metaphorically, the protreaty officials needed a tank but got a bicycle. Nevertheless, Sam quickly sensed their need for support and organized opposition to the unilateral legislation.

Despite tight time constraints in the spring of 1972, we carefully prepared analyses of the bill and information about the treaty for members of Congress and for likely allies in religious, world order, and environmental groups and among international lawyers and academics. We assembled witnesses to oppose the bill that Senator Lee Metcalf (Dem., Mont.) had introduced in the Senate and Representative Thomas Downing (Dem., N.Y.) into the House. The spring 1972 hearings were sponsored by congressional friends of the American Mining Congress, with only us organizing the opposition.

When we first entered the stately walnut-paneled House committee room in the Longworth building on May 25, 1972, we felt green and shaky. Who were we, fresh from a Virginia mountainside, to challenge some of America's leading corporations and such respected senators as Metcalf and Henry Jackson (Dem., Wash.)?

The witnesses on our side included John Logue, who said, "The Deep Seabed bill is a clear violation of the General Assembly Resolution on the common heritage of mankind of December 17, 1970." Another witness, Charles Clusen of the Sierra Club, saw the treaty as important, if not essential, environmentally, and feared that the legislation would jeopardize it.

In his testimony Sam was most concerned about the possible aborting of the negotiating process and stressed the broader U.S. interest in protecting and enhancing navigation, fishing, marine science, the environment, and North-South relations. By establishing new mechanisms for dispute settlement, he argued, a new law-of-the-sea treaty could lead to a more peaceful world order. Sam also argued that America's international reputation could be damaged by suspicions that the United States was appropriating for itself in Washington what it was negotiating for all at the United Nations: the right to mine the seabed, which the United States had never owned and never claimed.

Sam pointed out that "rights granted in the bill were not like high seas fishing rights where all may fish in the same area. An exclusive development right amounts to an appropriation of a non-renewable resource." Unlike fishing, in other words, mining requires rights of many years duration and exclusive rights within a defined area. Such uncontested rights could derive only from an agreement among owners of the property, the international community.

We hoped that a new law of the sea treaty would ensure those rights. We never opposed what we saw as the mining industry's legitimate needs. We did criticize its timing and many of its arguments, and we feared that greed could outrun need. Our research had persuaded us that the minerals in nodules were not scarce "strategic minerals," as the promining forces claimed, but were in most cases widely available from land-based sources.

Although industry stressed the economic advantages to the nation, Sam thought they were more likely only to aid the industry. There was one simple reason: if seabed minerals displaced land-based minerals from Canada, Zaire, and other nations, they would have fewer dollars to purchase U.S.-made products. He also questioned whether the industry's talk of lower prices to aid the consumer would materialize.

Columbia Law School professor Wolfgang Friedmann told the House committee that "any bill enacted by this country which would extend the jurisdiction beyond the continental margin would give rise to a series of counterclaims and effectively divide the oceans.... It would further erode what remains of freedom of the seas and provoke a variety of retaliatory measures which would greatly jeopardize the military, political, and economic interests of the United States."

Friedman's testimony had special meaning for us. Our friend Frank Porter Graham, a former university president, senator, and UN mediator, told us repeatedly in the last days of his life to work on law of the sea and to read Friedmann's compelling book, *The Future of the Oceans*, which warned of the rising tide of national claims against the sea (Friedman 1971). Friedmann's testimony was especially timely because, within four months, he was stabbed to death on the streets of New York.

The most important member of Congress who testified against the mining bill that spring was Sen. Alan Cranston (Dem., Calif.), whom Sam had gotten to know when both were active in the world federalist movement in the late 1940s and who always made time to consult with Sam on strategy on law of the sea. If the seabed mining bill were to become law, Cranston argued, "we will project the present international chaos onto the remaining 70 percent of the world's surface.... I believe we must find permanent structured regularized peacekeeping procedures."

Other members of Congress who helped us greatly in the early years were Senators Clifford Case (Rep., N.J.), Hubert Humphrey (Dem., Minn.), and Claiborne Pell (Dem., R.I.), and Representatives Donald Fraser (Dem., Minn.) and Berkley Bedell (Dem., Iowa). All were staunch internationalists who believed that the United States should nurture international law and global institutions. And all had legislative assistants who worked closely with Sam and me in the continuing struggle to prevent the passage of unilateral legislation.

With the possible exception of Senator Cranston, Representative Fraser was our closest friend and ally on Capitol Hill from 1972 until 1978, when he gave up his seat to make an unsuccessful bid for the Senate. As chair of the House International Relations Committee's

Subcommittee on International Organizations and Movements, he combined Midwestern practicality with a broad view of U.S. interests and insight into the UN and its problems. He also chose highly competent staff members. The first one we worked with, Robert Boettcher, attended early law-of-the-sea preparatory sessions and then gave us valuable detail and insight into the negotiating process. Boettcher also asked the Congressional Research Service and the Library of Congress to do specialized studies that ended up benefiting all of us on the protreaty side. These studies helped balance the promining reports that emanated from the Senate Interior and Insular Affairs Committee. Finally, Boettcher frequently called Sam for advice for Fraser's speeches and moves.

Once when I spoke in Minneapolis and appeared on a panel with Representative Fraser, I found OEP's most recent publication in two places: the notebook he was speaking from and in the hands of an editorial writer for the *Minneapolis Star and Tribune* who asked me to brief him at lunch.

These are some of the people whose need for our information and viewpoint helped ease the pain of the brush-offs, the frustrations, the occasional perfidy, and, above all, our inability to sufficiently expand a beachhead on Capitol Hill that might have prevented the Reagan administration's withdrawal from the treaty process. They helped make us forget that funds were lacking for the kind of campaign the well-heeled supporters of the seabed mining legislation could undertake.

The public-interest opposition was low on clout, but at least it was opposition. The officials who coordinated the protreaty effort were grateful. One day during these early years when Sam was visiting the State Department, Myron Nordquist proudly displayed to Sam 150 protreaty, antimining legislation letters to the president that were on his desk for reply. "For an issue where no American boys or pocketbooks are imminently threatened," he told Sam, "that is a flood."

In his final 1972 testimony, the chief U.S. delegate to the preparatory conference, John R. Stevenson, announced that the administration's decision for or against the seabed mining legislation would be delayed until the following year. When the decision was made, it was against the legislation on the grounds that it could harm the negotiations and preempt the conference. The administration's support for the position we were advocating gave us time to assemble a loose coalition and develop our own methods and style, knowing well that the struggle over seabed mining legislation was likely to escalate.

Supporting the Conference

During 1972 and 1973 the United States Committee for the Oceans became the spark for preparing and disseminating information on Capitol Hill by Sam, by his student assistants, and by a loose coalition of religious, world order, and environmental groups that I helped organize. We networked on this issue with other generally liberal internationalist groups, including the United Methodists, the National Council of Churches, the United Nations Association, the United World Federalists, the Federation of American Scientists, and the American Association of University Women. Among environmental groups, the Sierra Club was a standout. The earliest Catholic support came from the Center for Concern, a Washington-based group that focused on North-South problems; soon some members of the Maryknolls and a network of Catholic sisters became interested in the legislation as a social justice issue.

Unable to develop personal contacts with all 535 members of Congress, Sam and his associates on the U.S. Committee focused on staffs of the relevant committees for permission to testify; on proposals for drafting or revising substitute legislation on seabed mining and fishing; and on the sharing of information from our three arenas of activity—Congress, the administration, and the conference. Sam and his allies in Congress followed a well-known legislative technique of slowing legislation by lining up committees to request its referral to them. Because Congress reorganizes itself every two years and thus in many ways begins anew, this technique can be used to bottle up nonurgent bills for years. Thus seabed mining bills that had first been introduced in the Senate Interior Committee in 1971 spread to the House Merchant Marine and Fisheries Committee in 1972, and to many more general foreign policy, commerce, and environmental committees and subcommittees in future years. In fact, in the Ninety-fifth Congress (1977–78), treaty-related legislation was heard by six committees in sixteen separate hearings; fortunately from our standpoint, nothing of significance was passed during that session.

In the fall of 1972 Professors Sohn and Logue advised us to develop a positive initiative on behalf of law of the sea while we continued to oppose passage of the seabed mining legislation. They felt it was not enough "to be against something." Following up on this suggestion, Sam worked with Representative Fraser and especially with his aide, Robert Boettcher, to develop a resolution supporting the five basic U.S. goals that Ambassador Stevenson had enumerated during recent hearings on the seabed mining bill.

These goals stemmed from the 1970 draft treaty, and included among other things freedom of navigation and orderly development of (included "assured access to") deep seabed mineral resources. Sam spent long hours in December 1972 and January 1973 working on drafts of the resolution with Boettcher, with Myron Nordquist of the State Department and Leigh Ratiner of the Interior Department, and with experts on drafting legislation from the Congressional Research Service. Finally, the resolutions were introduced in February 1973 by Representative Fraser (House Resolution 216) and by Senator Pell (Senate Resolution 82).

Strong Defense Department support gave the resolution a big boost, as did the fact that the resolution was nonbinding and thus did not commit the Congress to support or to oppose any provision of a negotiated treaty. Backing from the Defense Department kept the resolution from being labeled "liberal" or "internationalist," and meant that staunch conservatives including Senators John Stennis (Dem., Miss.) and Strom Thurmond (Rep., S.C.) supported it. So did former secretary of state Dean Rusk, a believer in international law, who testified that the choice facing the United States was "common interest" or "conflict." The most important supporters were Nixon administration officials and the liberal internationalists in Congress, led by Fraser and John Seiberling (Dem., Ohio) in the House and Pell and Cranston in the Senate.

Although our support was not needed to pass the resolution, Sam and I lined up several liberal NGOs to testify on its behalf, including J. Elliot Corbett from the Washington office of the United Methodists and Anne Mason from the Center for Concern. We and our associates in the NGO community were thrilled when the House passed the resolution by a vote of 303–52 on April 2, and when the Senate approved it overwhelmingly by voice vote on July 9.[2] After the Senate vote Ed Snyder, Sam, and I celebrated in the gallery. "You can chalk that up to a good day's work," Snyder commented. Sam answered, "Representative Fraser made it happen."

---

2. The text of Senate Resolution 82 (like the corresponding House version) that endorsed the goals of President Nixon's oceans policy may be found in *Congressional Record—Senate* 119:18: 22812-13.

Opposing Two-Hundred-Mile Fishing Bills

Throughout the remainder of the Nixon and Ford administrations, Sam and his allies in Congress and in the NGO community fought mining industry efforts to pass unilateral mining legislation. Sam repeatedly testified before congressional committees on the issue, and lined up other religious, environmental, and liberal internationalist NGOs to do so as well. But the NGO effort was insignificant compared with two main reasons why the legislation did not pass: first, the administration consistently opposed unilateral U.S. legislation, arguing that the negotiators should be given time to work out a negotiated solution; and second, relatively few members of Congress were convinced by the mining corporations' argument that unilateral legislation was urgently needed to protect U.S. interests. It thus seems likely that the legislation would not have been passed—or if passed, not enacted over the president's almost certain veto—even if Sam and his associates had not been leading the NGO opposition to it from 1972 through 1976.

With the mining legislation contained for the present, the spotlight shifted to conflicts over fishing. Two NATO members, Iceland and Britain, were clashing in a "cod war" in the north Atlantic. Arrests of U.S.-owned tuna boats off western South America and the U.S. government's payment of their fines continued, as did conflicts over U.S. fishing for shrimp off the coasts of Brazil and Guyana. American fishing towns in the Northeast, the Pacific Northwest, and in Alaska stirred with demands to "get the foreigners out" of nearby waters. The United States should enact legislation for a two-hundred-mile fishing jurisdiction, many fishermen demanded. Aroused citizens blamed foreign vessels, especially the large eastern European factory fishing ships, for the depletion of stocks and the alleged destruction of their gear. Many sports fishermen joined the antiforeign outburst.

Pitted against the coastal fishermen were tuna and shrimp fleets that scoured the Latin waters. They were less numerous than the coastal fishermen, but they had larger overall revenues. They opposed a U.S. two-hundred-mile limit because, if the United States passed such a law, American officials no longer could complain about the two-hundred-mile fishing jurisdictions that several South American nations had proclaimed unilaterally.

And so the congressional battle was joined over the two-hundred-mile fishing bills that began to be taken seriously in 1973 and 1974.

This legislation had the potent backing of Sen. Warren Magnuson (Dem., Wash.), chair of the Senate Commerce Committee. Senators with no fishing concerns beyond the brook trout dared not cross Magnuson because they had to come to the Commerce Committee for funding for small business loans and other economic development projects.

Officially, the Nixon and Ford administrations opposed the legislation. But without consistent leadership at the top, different agencies often sent out contradictory signals. Some of the fisheries' staff at the Commerce Department, for example, hinted that the administration would approve a two-hundred-mile bill.

State Department officials whom we worked with were determined to abide by a treaty that the United States had ratified: the 1958 Geneva Convention on Fishing on the High Seas. How, these officials argued, could the United States discourage anarchic unilateralism if it flouted this agreement? If we denied other nations' rights to fish in 2.8 million square miles of ocean off our shores, how could we expect them not to deny our ocean users' rights near theirs?

Opposition to a two-hundred-mile bill was widespread in the State, Defense, and Commerce Departments. If the bill was enacted, many officials feared, the United States would lose bargaining leverage at UNCLOS III. U.S. negotiators hoped to receive improved provisions on navigation and scientific research within two hundred miles in exchange for our granting coastal nations the two-hundred-mile jurisdiction over resources that they insisted upon. State Department officials, including John Norton Moore and Myron Nordquist, repeatedly gave these and other reasons in congressional testimony against the bill.

Sam and his associates in the U.S. Committee and in the Methodist Project joined U.S. officials interested in law of the sea and shrimp and tuna interests in leading the fight against the two-hundred-mile bill. A key figure in delaying the legislation was Rep. Leonor K. Sullivan of landlocked St. Louis, the chair of the House Merchant Marine and Fisheries Committee.

Our hard-working interns, Brad and Hannah Rishel and John McLean, wondered how a nonseacoast representative could be chairing that committee, which normally is dominated by representatives whose coastal districts are home to local or distant-water fleets. We told them that the giant Ralston Purina Company was centered in St. Louis, and that several of its pet foods contain the scrappy parts of tuna. Sullivan opposed the legislation because she feared that it would legitimize the restrictions and fines that Ecuador and Peru placed on U.S. tuna fleets. In our office Sullivan acquired the nickname "Cat Chow Queen," which in no way diminished our gratitude for her position or our respect for her legislative skills. We also appreciated the fact that Mrs. Sullivan and Rep. Charles Dingell (Dem., Mich.) introduced bills that Sam and State Department officials had worked up that sought to increase conservation while keeping U.S. policy within the terms of the 1958 convention. These bills were opposed in the House by two-hundred-mile bills introduced by Rep. Gerry Studds (Dem., Mass.) and others.

During 1974, an election year in which the events leading up to President Nixon's resignation in August held center stage, opponents of the two-hundred-mile bill kept it from passing, despite increased pressure from coastal fishing interests and their supporters in Congress. But pressure built in 1975: in May the House Merchant Marine and Fisheries Committee voted out the two-hundred-mile bill, and the full House passed it on October 9.

Earning gratitude from his friends in the State Department, Sam kept up the fight against the bill in the Senate, working especially with Senators Cranston and Marvin Griffin (Rep., Mich.) on alternatives to the two-hundred-mile bill. The following excerpt from a letter I wrote to William Fischer on December 16, 1975, suggests both Sam's hard work and the difficulties we faced:

> Sam has been working on the Hill. He saw Senator Cranston at 11:30 this morning trying to get him to circulate a Dear Colleague letter to the other 99 Senators which would answer extreme statements by the proponents of the [two-hundred-mile] bill.... Cranston said that Magnuson was angry and that he wanted to lay low and not offend him. Sam thinks that Sen. [Mike] Gravel [Dem., Alaska] can delay the bill until after Christmas.... Two State Department officials came to the strategy meeting of our side the other day, and are leaning on Sam so heavily it's almost pathetic. (Miriam Levering 1975a)

Already in deep trouble, our cause became doomed when President Ford, facing pressure from New Hampshire fishermen/voters and a serious challenge from Ronald Reagan for the Republican nomination, decided in January 1976 to switch his position and publicly support the two-hundred-mile bill. The Senate then approved the bill on January 28 by a 72–19 vote. Sam and I took some consolation because the major alternative, the Cranston-Griffin amendment, lost by a much smaller margin, 58–37.

Sam felt exhilarated by the struggle: he had learned much about ocean fishing problems and conflicts and about the legislative process, and he had used the battle to stress the importance of an eventual law of the sea treaty. He was grateful to his colleagues—NGO, Capitol Hill, and administration—and pleased to receive a letter of thanks from Nordquist at the State Department in early February. "No one was more effective in opposing the legislation than you," Nordquist wrote.

Sam, in turn, was especially grateful to one of his volunteers, a Wooster College student named John McLean. All night long he threw up sacks of mail in the House Post Office to support himself. After breakfast he came to work for the U.S. Committee, researching, writing, poring over bills with his eyes half open. Our volunteers repeatedly impressed and inspired us.

Fortunately, the two-hundred-mile fishing bill was not a huge setback for the conference, as a unilateral mining bill almost certainly would have been, because the negotiators clearly were moving toward a two-hundred-mile exclusive economic zone for coastal nations. On the other hand, U.S. negotiators had not finalized the nature of that zone; they especially worried that it would become as much under national control as the twelve-mile territorial sea. Their fears seemed confirmed when, at the next session of the conference, the Soviets joined the large group of Third World delegates who wanted to restrict freedom of research within two hundred miles of shore.

Supporting the State Department

Besides working with like-minded people on Capitol Hill and in the NGO community on seabed mining and fishing legislation, Sam and I also developed close contacts in the administration during the early years of our work on law of the sea. Still relatively new to Washington, we had little impact on actual administration policy; but by working hard to support the administration's (and our own) positions in Congress and in meetings of the Public Advisory Committee to the Joint Interagency Task Force on Law of the Sea (hereinafter called the Public Advisory Committee), we built up a reservoir of gratitude and respect that benefited us in later years.

Sam especially enjoyed working with protreaty State Department officials on law of the sea: Myron Nordquist, Bernard Oxman, George Taft, and others. Strong supporters of a negotiated treaty, they provided Sam and me with much of the information that we used for our work on Capitol Hill and our efforts to counter antitreaty forces in the Public Advisory Committee and in Washington generally. I included much of this information in *Sea Breezes,* the U.S. Committee's newsletter that I edited and largely wrote during these years.

It almost certainly was gratitude for Sam's protreaty work on Capitol Hill that led several State Department officials and Leigh Ratiner of the Interior Department to recommend him for membership on the Public Advisory Committee in late 1972 and early 1973. At their urging, Sam twice applied for membership, and was successful early in 1973 on the second try. U.S. policy on ocean law was made by the Joint Interagency Task Force of the National Security Council, which included undersecretaries of State (who served as chair), Defense, Treasury, Interior, Commerce, and Transportation. This body wrote the delegation's instructions for each session of the conference.

With the need for eventual senatorial approval in mind, Congress in 1971 established the Public Advisory Committee, which was to include representatives of major ocean interests. In turn, there were subcommittees on which the relevant industries were well represented that dealt with such areas as petroleum, hard minerals, fishing, and shipping. There also were subcommittees on scientific research, international law, and the environment. The Public Advisory Committee met before and after each session of the conference, debated and discussed U.S. policy, and became acquainted with each other and the key U.S. negotiators. Committee members were allowed to serve on the U.S. delegation and had access to some classified material.

Ratiner and protreaty State Department officials wanted Sam on the Advisory Committee to balance some of the antitreaty members, especially those who represented seabed mining interests. Protreaty officials appreciated the work that Sam was doing to build support for the treaty in Congress, and wanted him at the Public Advisory Committee meetings as an advocate of ocean law over unilateral action. When Sam spoke at the committee meetings, he often stressed the need to take a long-term view of U.S. interests, arguing that American interests would be protected better by a treaty than by unilateral action.

Although not a member of the Public Advisory Committee, I was invited to attend the receptions that marked an evening break in the day-and-a-half-long sessions. These occasions helped our colleagues become people to us, with family problems and children to educate. Although we disagreed with Conrad Welling of Lockheed on the seabed mining legislation, for example, we shared his concern for the safety of his daughter, a U.S. Navy test pilot. Personal contacts Sam and I developed at the receptions strengthened our more businesslike relationships with decision makers and with other NGO and industry representatives in Washington and at the conference.

Serving on the Public Advisory Committee made Sam keenly aware of splits within the administration on ocean policy. He learned details of the split between Defense and Interior over the breadth of the continental shelf and margin. Defense feared "creeping jurisdiction," with its potential to impede military and commercial navigation. Navy officials also feared that coastal-state control might severely limit their ability to place submarine-tracking systems on portions of the ocean floor that nations claimed. Defense thus supported coastal-state control of the seabed only out to the two-hundred-meter isobath (that is, to the point at which the continental shelf is two hundred meters below the surface of the ocean). Interior, strongly supported by most oil companies, wanted to have all recoverable energy resources under the ocean floor controlled by the coastal states. Except for Ratiner, most Interior officials also wanted their agency—not an international authority—to license U.S. deep-seabed miners.

In 1973–74, Sam and his allies at the State Department found the Treasury Department to be the agency most hostile to the treaty process. To him that department had a partially understandable but overly narrow concept of the U.S. national interest and an insensitivity to the current international climate. It was 1973, not 1945, when the U.S. dominated global decision making. No longer could the United States, acting alone or with a few other nations, dictate the use of resources in a nonnational area or the rules that governed more than half of the earth's surface. Officials at Treasury either did not accept the concept of shared ownership or they did not agree that decision making has to be shared when ownership is shared.

Specifically, the Treasury Department under secretaries William Simon and George Shultz opposed the idea of the licensing of U.S. seabed-mining corporations by an international seabed authority and the payment of a share of the profits to an international organization instead of to the U.S. government. Treasury was unconvinced that it should allow tax credits for such payments, as is the customary practice when foreign governments tax American corporations. The intensity of Treasury's opposition to revenue sharing (a central part of the 1970 U.S. draft treaty) upset the State Department officials with whom we worked. They suspected a rigid free-market ideology that opposed any concessions to Third World diplomats.

Our colleague, William Fischer Jr., confirmed Shultz's position personally when he followed the Treasury Secretary to an elevator after Shultz had spoken at a dinner sponsored by the World Affairs Council of Philadelphia. "What about law of the sea?" Fischer inquired. "I'm against it," Shultz responded and stepped onto the elevator.

Had a president or secretary of state or defense considered law of the sea a high priority and intervened on its behalf, the interagency conflicts of this early period might have been more easily contained and the U.S. delegation headed by John R. Stevenson might have been able to speak with one voice. But neither President Nixon nor President Ford, nor the secretaries of state and defense who served under them, provided this leadership.

Secretary of State Henry Kissinger took the most personal interest of the pre-Reagan secretaries of state. Kissinger was the only secretary of state to appear at a conference session—in New York in 1976. But according to Carlyle Maw, Kissinger's legal adviser and friend at the State Department, Kissinger dreamed of calling a few foreign ministers together to put together the treaty. "That won't work," Maw remembered telling him.

Awed by all the high officials and business executives in attendance, Sam spoke sparingly at most of the early meetings of the Advisory Committee. But in the meetings of January 10–12, 1974, John E. Flipse, representing the Deepsea Ventures mining firm, accused Sam of getting information for the U.S. Committee's newsletter from classified documents that the State Department had allegedly leaked to him. Sam responded that he had gotten details of the split between State and Treasury from interviews with Treasury officials, not from the State Department.

This opening gave Sam an opportunity to support the position of the State and Defense Departments and to emphasize the U.S. interests in international cooperation, navigation, stability of investments, dispute settlement, environmental protection, and other matters that could be advanced through the treaty process, not through unilateral legislation or total refusal to share revenues produced in seabed mining. As he watched some heads nod around the room, he hoped that what he perceived as a Treasury effort to block further U.S. participation in the negotiations had miscarried.

Much to Sam's and my satisfaction, U.S. policy became more consistent—and more broadly protreaty—when Jimmy Carter became president in January 1977.

4

Gaining Legitimacy at the United Nations Conference

IT BECAME CLEAR TO US from our Washington struggles to protect the conference from unilateral acts that the negotiations needed help. Speedy resolution of issues was needed if unilateral measures and tensions between developed and developing nations were not to overwhelm the conference. Fortunately, most governments at the time viewed the UN as a viable if not optimal instrument for improving international relations. There also was widespread support, at least in principle, for protecting the environment and trying to improve living standards in developing countries. The easing of the cold war also contributed to the widespread optimism of the early 1970s. Although Sam and I knew that the negotiations might well fail, we also believed that there was a reasonable chance that they might succeed. For anarchy in the oceans was in no nation's long-term interest.

During the fall of 1972 Sam and I met in New York with some UN-based diplomats and NGOs to explain what we were trying to do in Washington and to learn in what ways we might be helpful in the negotiations. Several diplomats thanked Sam for opposing unilateral U.S. legislation at a meeting at Quaker House near the UN, which became the center for our work when the conference met in New York in future years. I made contact with several NGOs in New York, notably Patricia Scharlin-Rambach, director of the Sierra Club's Office of International Environmental Affairs, whose organization sponsored a project jointly with us at the Caracas session in the summer of 1974.

My initial contact with the preliminary negotiations came when I attended the last meeting of the preparatory conference from July 4 through August 7, 1973. I went to Geneva with two main goals: first, to learn all I could firsthand to improve our effectiveness in all arenas; and second, to try to persuade negotiators to reserve some revenues from the seabed for the international community before coastal states' appropriation of seabed resources foreclosed this hope. In other words, we urged not only that revenues be shared from beyond the proposed two-hundred-mile limit for national control of resources (mainly from deep seabed mining of nickel and other hard minerals), but also that revenues be shared as well from more easily exploitable resources from the continental shelf (mainly oil and natural gas) beyond the twelve-mile territorial limit. The 1970 U.S. draft treaty had favored this approach, and I rather innocently thought that I might be able to build support for it.

My young American colleague Jim Bridgman and I took our message from one government's office, or mission, to another. Representatives of landlocked nations agreed with us but felt themselves outnumbered and did not wish to jeopardize their chances of receiving access to the sea through their coastal neighbors, plus some fishing rights in coastal waters, by pushing their neighbors too hard on revenue sharing from oil and natural gas.

Even the Scandinavians, who often were generous in supporting international institutions and Third World development, were cool to our ideas. A lively Dane said the proposal was "not in the cards," noting that because of Denmark's ownership of Greenland "we've joined the island Mafia" that were appropriating as much of the oceans as possible. A Norwegian official was even more blunt: "If the taxpayers of Norway wanted to share oil revenue with the UN and developing countries, they would have told us."

My month in Geneva that summer was indeed a learning experience. From the interviews I began to realize what strong forces national self-interest and regional alliances would be in the negotiations. I learned that the desire of many NGOs, including Sam and me, to achieve a treaty that would shift substantial wealth from the world's richer nations to poorer ones was utopian. As the secretary of the conference, David Hall, responded after one of my pitches for revenue sharing, "Mrs. Levering, if we can get any treaty at all, I will be happy." I also learned from runarounds on appointments and from the large number of sessions closed to NGOs that many diplomats and UN officials were cool to NGO participation in the conference. "We toil under greater difficulties than in Washington," I wrote somberly to Sam on July 15. But he and I did not let these difficulties dampen our determination to contribute positively to the conference.

Learning from Caracas

Besides opposing unilateral legislation in Washington and building support for our position among other NGOs in Washington and New York,[1] Sam and I spent considerable time in late 1973 and early 1974 trying to raise money for our work. We were able to raise roughly $40,000, mainly from foundations, that enabled us, in conjunction with other NGOs, to plan activities for the second session—the first substantive one—of UNCLOS III. This session, which met in Caracas, Venezuela, from June 20 to August 29 of 1974, attracted several thousand delegates from more than one hundred nations, plus several hundred journalists and NGOs largely from North America and western Europe. While progress was made at Caracas on a few issues, such as the two-hundred-mile economic zone, the session primarily demonstrated how much hard work lay ahead before this large and diverse gathering could even begin to achieve broad agreement on a treaty.

If energy and enthusiasm are proper measures of NGO effectiveness, then our efforts were successful. Working with the Sierra Club and, more informally, with several other NGO representatives, Sam and an energetic young staffer, Bill Sargent, organized three panel discussions on science, technology, and the oceans in July that together attracted more than four hundred delegates plus journalists and NGOs. The specific themes of the seminars were the ocean environment (featuring the well-known writer Thor Heyerdahl, among others), seabed minerals, and fisheries.

Keenly interested in obtaining media coverage for environmentalist thinking, Sargent also organized several news conferences and appeared on at least one local television program. Sam and I preferred

---

1. For example, Sam and I and John Logue reported on the recent meetings in Geneva at an NGO working group on law of the sea in New York on September 25. The following organizations were represented at the meeting: American Friends Service Committee, Committee to Study the Organization of Peace, National Audubon Society, Natural Resources Defense Council, Ocean Education Project, Quaker House, Sierra Club Office of International Affairs, U.S. Committee for the Oceans, World Association of World Federalists, and World Order Institute.

Part of the OEP team at Caracas in 1974. From left: Jim Bridgman, Miriam Levering, Arthur Paterson, and Wendy Witherspoon. Courtesy of Miriam Levering.

the more traditional approach of getting to know delegates—especially ones who shared our world order perspective—and meeting with other NGOs to discuss how to be most effective.

Another of our young associates, Arthur Paterson, helped us become acquainted with the delegate who would become our best friend among non-American diplomats, Ambassador Tommy T. B. Koh of Singapore. Early in the session Arthur became friends with Geoffrey Yu of Koh's staff. Because Arthur faced the unpleasant prospect of sleeping on the floor of an NGO apartment for the duration of the eight-week session, Yu invited Arthur to share his apartment. Through Yu, Arthur got to know Koh. Paterson commented later on what he and other NGOs working with us learned at Caracas:

> Since NGOs had largely been more gadflies than participants in most intergovernmental political conferences [up to that time], we found it difficult to follow the negotiations, day by day, without beginning to cultivate long-term relationships with negotiators. International journalists were having the same difficulty covering the conference, though few were present in Caracas. NGOs and several reporters, in

particular Canadian, met regularly after about the third week of the conference to share information and gossip. Corralling delegates in corridors and meeting with them over endless cups of coffee and cigarette smoke proved our baptism to one style of NGO participation. Without current information on the status of the negotiations, we could neither garner information for ourselves and others nor could we learn to design programs that were directly relevant to the current status of the negotiations. (Paterson 1984)

In thinking that fall about what we had done at Caracas and what we needed to do at upcoming sessions, we realized that the kinds of activities Paterson described—getting to know delegates and trying to design programs that could assist them in their work—probably was our most important contribution. Overall, however, we sensed that much of our work at the conference had been amateurish and should not be repeated at future conference sessions. We had put too much time and effort into pushing internationalist positions such as revenue sharing from oil and seeking ephemeral media coverage of our views, and not enough time listening to delegates and asking them how, if at all, we could be helpful to the negotiations. We realized as well that most delegates were more interested in learning about ocean resources and how they might be regulated than they were in listening to fervent environmentalists such as Heyerdahl or committed advocates of world order and justice such as Logue and ourselves. Moreover, friendly U.S. diplomats told us that we were hurting our cause by distributing protreaty pamphlets that we had written to influence the political debate in Washington, but that were not relevant to foreign diplomats' concerns.

In short, we were beginning to learn that, if we wanted to be taken seriously at the conference, good judgment and usefulness to the delegates would be more important than good intentions and zeal on behalf of preferred outcomes. But the well-attended general seminars that we organized at Caracas at least allowed us to become acquainted with many delegates besides the ones I had met the previous year in Geneva. These personal contacts gave us something upon which we could build.

Achieving Greater Legitimacy at the Conference

Fortune smiled on our work during the six and one-half months between the end of the session in Caracas (August 29, 1974) and the start of the third session in Geneva (March 17, 1975). For one thing, the person who was to become the single most effective member of the Neptune Group in the UN arena, Lee Kimball, began working for OEP as a consultant in the summer of 1974 and continued working full-time until the end of the Geneva session in May 1975. For another, money and coworkers seemed easier to come by than they had been earlier: not only did foundations pitch in to help with expenses, but also the Women's Division of the United Methodist Church began to make larger contributions. Besides financial help, which included $2,000 to publish our pamphlet *The Oceans: Time for Decision* in January 1975 and another $2,500 to assist with publishing costs at Geneva, the Methodist Women decided to fund a six-month project on law of the sea (a project that subsequently became ongoing) that included sending two members of their staff, Joyce Hamlin and Barbara Weaver, to assist us in Geneva. With the arrival of Kimball, the Methodists' commitment, and my lining up several volunteers to help with our work in Geneva, the Neptune Group thus emerged by late 1974, several months before the first issue of *Neptune* came off the press in Geneva in March 1975.

Kimball threw herself wholeheartedly into her work with me at OEP that fall and winter. Seeking funding for our work in general and for her salary in particular, she wrote at least eighteen grant proposals to foundations. She built on our earlier contacts with reporters and editors at the *Washington Post* and other newspapers. She worked with Sam and me on a variety of in-house publications and articles for other media. And she largely organized a conference in New York on January 23–24, "Law of the Sea: Resource Questions and Economic Development." This conference, cosponsored by OEP and Project Dialogue of the Carnegie Endowment for International Peace, featured twenty-three speakers, mostly diplomats and experts on ocean issues from industry and academia. Kimball wryly described some of the difficulties involved in lining up speakers in a January 17 letter:

> Our seminars next week in New York look like they're finally going to happen. I've racked up more telephone hours than any operator ever thought of... but the hardest part is deciphering the accents when they say Ambassador so-and-so or Mr. so-and-so is calling and I have to quickly dredge up from my slow recall which particular mission at the U.N. he is associated with so as not to have to ask and offend his pride. (Kimball 1975b)

The most important work that Lee, the United Methodists, and I did that winter was to plan OEP/UMLSP activities for the upcoming

Geneva session and to line up people to help us carry them out. Lee and I sought advice on how we could be most helpful from diplomats we knew and from NGOs in the United States and in Geneva. Aware that communication had been poor among the many delegates and others who had gathered in Caracas and that a newspaper called *Pan* (bread) had made a contribution at the recent World Food Conference in Rome, we eventually decided to publish a newspaper that would appear roughly every week to disseminate information and opinion. Responding on January 21 to a letter from Lee, whom Sam and I had asked to organize and edit the newspaper, Rene Wadlow, a World Association of World Federalists representative in Geneva, offered both encouragement and a suggested name: "Your idea of a newspaper for the conference sounds like an exciting one. I think after the god Pan for food the most obvious is Neptune for the sea" (Wadlow 1975). Thus a fellow NGO representative suggested the name for the newspaper—a name that quickly became associated with the diverse and frequently changing group of mostly young people, led by Kimball and me, who published it over the next seven years.

Our other main effort that winter was Sam's work to speed the search for a compromise between the developed and developing countries on deep seabed mining. Concerned about the pressures building in Congress to pass unilateral U.S. legislation, Sam advocated a compromise that would combine international control of the deep seabed with assured access to seabed minerals for mining companies that met fair international criteria. Sam distributed a draft of his proposal at a conference on ocean law in Miami in early January, and urged representatives he met from developing nations to compromise at the Geneva session. Sam also sent copies of his proposal to the deputy head of the U.S. delegation, John Norton Moore, and to diplomats from around the world who planned to attend the upcoming session. Wanting to keep his freedom of action as an NGO, Sam declined Moore's invitation to be a member of the U.S. delegation at Geneva.

The Geneva session (March 17–May 9, 1975) marked the emergence of what the delegates began to call the "Neptune Group" as a significant player at the conference, largely because of the publication there of six issues of *Neptune*. But the extent of our impact at this session should not be exaggerated, for we were still inexperienced and overly idealistic, and we did not know how to help the delegates as much as we would in later years. Still, Arvid Pardo commented—

rightly, I believe—that, because of *Neptune,* we had made the greatest NGO contribution at Geneva. His comment encouraged us, as did the request of many delegates that we continue to publish *Neptune* at future sessions. As one of our staffers, Jim Orr, recalled: "We felt we were about the one written vehicle—since there were no newspapers really covering this [the session] on a daily basis—the one group that could bring disparate facts together and make points and challenge statements and conventional wisdom" (Orr 1989).

Overall, the Neptune team was much larger at Geneva than it had been at Caracas: something like eight to nine people at any given time, compared with an average of three or four the previous summer. Kimball, Jim Bridgman, and I—all from OEP—were there for the entire session. We were aided in the early weeks by my sister Carol Lindsey and by Joyce Hamlin and Barbara Weaver from the United Methodist Project, and by Sam toward the end of the session. A British Quaker and political scientist with expertise on law of the sea, Roderick Ogley of the University of Sussex, contributed ideas and articles to *Neptune.* But it was a motley crew of intelligent, idealistic young people, including Bridgman, who did the most to help Kimball and me write and publish the newspaper: Jim Orr, a friend of Lee's from graduate school days at SAIS; Dale Andrew, a Quaker friend of mine who, together with Orr and Kimball, had taken Prof. Ann Hollick's oceans policy course at SAIS; Carolyn White, who arrived unexpectedly and did the illustrations for the early issues; and John Diamante, a member of the Oceanic Society who had the only significant experience in journalism in the entire group. Without Diamante's assistance in layout, the first couple issues would have looked very amateurish.

Everything connected with writing and publishing the early issues of *Neptune* was difficult. At times delegates would not show up for scheduled interviews; staffers' feelings were hurt as stories were extensively rewritten; putting the newspaper together required all-night sessions of typing followed by cutting and pasting; and taking care of mundane details often was time-consuming in a French-speaking city that none of us knew well.

The biggest problem, however, was a two-sided conflict involving John Diamante. One side involved *Neptune*'s editorial direction. Lee and I wanted to keep *Neptune* as factual and informative as possible—that is, to emphasize usefulness to delegates and others at the conference—whereas John leaned toward advocacy and toward seeking attention from other news media for our personal views (Kimball 1995). The conflict's other side was a struggle for power. "Lee and Miriam thought immediately that his [Diamante's] real purpose here was to take over the paper and oust Lee as editor," Jim Bridgman noted in his diary. "He certainly had grand designs" (Bridgman diary n.d.). Commenting more generally, Joyce Hamlin recalled "late-night sessions and all kinds of personalities, some with horrendous egos, with Miriam to balance, to see that it got done before 4 a.m." (Hamlin 1989).

Although I appreciated John's contribution to the first and second issues, the *Neptune* operation ran more smoothly after he left Geneva on April 4. Nor did I respond favorably when he asked the following winter to join the Neptune Group at the New York session.

The big question for us was, what would the UN officials and delegates think of *Neptune*? The UN officials who oversaw the conference tended to be skeptical at first, because they remembered that the NGO environmental groups—and their newspaper—had contributed to the frenzied atmosphere at the environmental conference in Stockholm in 1972. At Stockholm, NGO representatives frequently had pointed accusing fingers at governments and had made sensational charges to attract media coverage. UN officials and delegates understandably did not want to have this recent history repeat itself at UNCLOS III. Thus, for the first issue UN officials ruled that we could place copies of *Neptune* only on the documents table. They also insisted that we remove any suggestion that the first issue of *Neptune* was official by scratching out by hand on each copy the UN logo and the UNCLOS III logo that we unthinkingly had placed on the masthead.

The primary concern of UN officials and delegates, however, was that we not publicize information about particular governments' positions that had been stated in confidence in closed meetings. "Finger-pointing at positions deemed 'negative' by the observer was especially frowned upon," Kimball recalled (Kimball 1995). Satisfied that the Neptune Group respected confidentiality and sincerely sought to help the negotiations, UN officials eased their restrictions for subsequent issues, enabling us to distribute many of the five thousand copies in coffee bars and in reception areas.

The reaction of delegates is more difficult to judge. Some delegates were more favorably disposed to us—and to NGOs in general—than were others. Moreover, in the early years some delegates—especially from Third World and communist bloc countries—suspected that we might be funded by the CIA. Here our penny-pinching ways and religious connections helped: Quakers (and, to a lesser extent, Methodists) were widely known and respected in international forums, and we held our weekly seminars at Quaker House in Geneva.

Still, many delegates initially were wary about *Neptune*. "At first, people didn't want to even take it," Joyce Hamlin recalled. "We had trouble distributing it" (Hamlin 1989). But by the second and third issues—and thereafter—we observed delegates avidly reading *Neptune* both in the coffee shops and during speeches in the plenary sessions, and we repeatedly received compliments, criticisms, and suggestions of what we should include in future issues. Feeling "absolutely bushed" from the eighteen-hour workdays required to produce *Neptune*, Jim Orr was heartened by the positive reaction of many delegates: "I was very surprised at the favorable reaction of a lot of Third World delegates. Right away, after they had read the first issue, when they figured out who we were, people would stop us and say, 'Gee, this is really a valuable service that you're providing us.' It gave us the motivation to continue" (Orr 1989).

Jim Bridgman's diary entry for April 16 echoes Orr's comment about the delegates' acceptance of *Neptune:* "I spent the morning doing interviews. I kept getting extremely positive reactions to *Neptune*. Some people were already pushing for a sequel to *Neptune* at the next session" (Bridgman diary). Although we liked the compliments, even more gratifying was the sense that, through *Neptune* and through seminars, we might serve as a catalyst that could help facilitate ultimately successful negotiations. Bridgman's April 6 diary entry cites progress in achieving the group's desired "information role": "In committee II the landlocked states have started a new drive to gain recognition and acceptance of their ideas. *Neptune* has done some articles on the landlocked states' position and [Joseph] Warioba of Tanzania has called the paper biased against the coastal states. More maps and facts were asked for by the Secretariat. At least we seem to be playing our information role. It is exciting" (Bridgman diary).

Before long, delegates from developed and developing states were coming to us and giving us material for *Neptune*. I remember U.S. officials handing us copy, and then the West Germans. The Soviets gave Roderick Ogley material for his article on seabed mining (Ogley 1975). We quickly found that we were getting more articles than we had room to include, and thus had to disappoint several delegates. One official whose material we published was Leigh Ratiner, the chief

U.S. negotiator on deep seabed mining, who anonymously wrote an article to acquaint delegates from developing nations with the various types of joint ventures that could be used in mining contracts between the international seabed authority and private companies. Seeing joint ventures as a promising way to break the deadlock on seabed mining, Lee and I also published a signed article on joint ventures by Prof. George Kent of the University of Hawaii (Kent 1975).

Overall, the six issues of *Neptune* reflected the approach that we took to the negotiations in general: first, we wanted a workable, widely accepted treaty—that is, one that necessarily would have to be based on compromise; and second, we wanted one that would be as generous as possible to the poorer nations and to those that had short coastlines or were landlocked. The first goal gave us some credibility as realists with Western delegates and mining industry representatives, and the second increased our standing with delegates from the G77 Third World countries. Maintaining a balance between the two goals separated us from some of the more rigidly pro-Third World NGOs on law of the sea—notably Logue and Borgese. Borgese alienated many delegates by lobbying persistently at Geneva for a treaty that would transfer much wealth from the oceans from the richer to the poorer nations.[2]

Besides publishing *Neptune* and putting on weekly seminars for NGOs and journalists on issues facing the conference, we were pleased with the personal relationships we established with many delegates (which gave us access to inside information) and the individual briefings we gave to journalists from the United States, Great Britain, Japan, and other countries to help them understand what was happening at the conference. Led by Dale Andrew, we also prepared four reports on the conference for National Public Radio's main news program, "All Things Considered."

In retrospect, we made one major mistake at Geneva. In my desire for publicity for our group and for the "save the seas" viewpoint, before leaving Washington I had written the famous French ocean explorer and activist Jacques Cousteau to ask him if he would like us to schedule a news conference for him in Geneva. When he called at the last minute a couple months later, we hastily set up a news

---

2. Jim Bridgman noted at the time that Borgese "had made herself unpopular with many delegates because she had arrived at the conference as an outspoken NGO. . . . Many coastal state countries thought she was unrealistic and a bit crazy" (Bridgman diary, April, 15 1975).

At Geneva in 1975. From left: Lee Kimball, Jacques Cousteau, Carol Lindsey, and Miriam Levering. Courtesy of Lee Kimball.

conference on April 24. Stating that the oceans might well be dead within fifty years, Cousteau sharply criticized "inaction" on ocean pollution, "the technological and economic hegemony of the most developed nations," and the proposed two-hundred-mile exclusive economic zone for coastal nations.

At the time, we were delighted that Cousteau's comments were widely reported in the European press, and that our group had facilitated this coverage. But most delegates considered Cousteau's views—especially his opposition to the two-hundred-mile economic zone—extreme and impractical, and we soon realized that sponsoring him had hurt our reputation at the conference. When he asked us to set up another meeting with the press a couple years later, we declined.

Improving the Usefulness of the Group's Seminars

Sam, Barbara Weaver, and I had almost ten months after the close of the Geneva session in May 1975 to think about how the Neptune

Group could be most effective at the upcoming fourth session in New York (March 15–May 7, 1976). Much would depend on how much money we could raise. A big disappointment during these months was that neither we nor the United Methodists could afford to offer full-time employment to Lee Kimball, which meant that Lee's energy and expertise were not available to the Neptune team until we were able to scrape together a small stipend for the second New York session (August 2–September 17, 1976). But the Methodists were able to give Arthur Paterson a modest salary, primarily to edit six issues of a newsletter called *Soundings* that was distributed to a largely new Methodist constituency as well as to those on the ever-expanding OEP mailing list.

Even with Kimball as editor and sufficient money, however, we still might not have chosen to publish *Neptune* frequently during the New York sessions, as we had at Geneva. The main reason was that Louis Sohn, the Harvard law professor and U.S. negotiator whom we admired so highly, urged us in November 1975 to focus at the conference on facilitating private meetings rather than on publishing *Neptune*.

Like us, Sohn was painfully aware of the huge gap on seabed mining between western and Third World delegates, and feared that, as I wrote in my notes on a talk Sohn gave in Washington on November 7, "if the atmosphere is hardened, information won't be paid attention to." Sohn saw private meetings and seminars as the best way to ease the atmosphere and get the delegates talking seriously with one another. Sohn also advised us "to brief the media so that the public can appreciate the magnitude of the issues involved" (Miriam Levering 1975b).

Some of Barbara Weaver's contacts at the UN and one of the members of the Neptune Group in Geneva, Roderick Ogley, also were concerned that *Neptune* might harm the negotiations. We thus decided not to publish *Neptune* during the conference, but to put out an issue (number 7) that would be sent to delegates and others before the New York session, and another, wrap-up issue (number 8) that would appear in May as the session was drawing to a close.

In these issues and in number 9 published for the August–September session, we encouraged delegates to persist in their quest for a broadly acceptable treaty, and presented ideas that we hoped would move the negotiations forward. An innovation in number 9 was an "idea box" that contained numerous suggestions for the negotiators, many of which grew out of the seminars and off-the-record discussions that in 1976 superseded *Neptune* as our primary contribution to the conference.

Before proceeding to describe some of these meetings and our other work at the two sessions in 1976, I should point out that in many ways the conference had become stalemated on seabed mining by the close of the 1975 session in Geneva and then became further polarized during 1976. The basic problem was that the Western industrialized nations, led by the United States, insisted on guaranteed access to the minerals for their privately owned mining companies, whereas the developing G77 nations wanted control over whether, by whom, and under what conditions seabed mining could be undertaken plus a large share of any profits realized from seabed mining.

Western negotiators were willing to concede some of the profits and even, according to a compromise proposal offered by Secretary of State Henry Kissinger, to help an international entity that could be established under the treaty to engage in seabed mining alongside private companies. But Western leaders were unwilling to negotiate treaty provisions in which a majority of nations could block mining by firms from the technologically advanced states, and they were reluctant to make concrete promises of funding and mining technology for this new entity, the "Enterprise." Because seabed mining was a prospective rather than an existing industry, because existing international law offered few if any precedents in this area, and because Third World diplomats saw the issue as a way to advance the goals of their hoped-for "new international economic order," the negotiations on this issue at UNCLOS III always were highly complicated and frequently were acrimonious as well.

To abbreviate a long story, Leigh Ratiner, the chief U.S. negotiator on seabed mining, met regularly at Geneva with Christopher Pinto of Sri Lanka to try to work out a compromise on seabed mining acceptable to both sides. Ratiner and Pinto made real progress toward a possible deal; indeed, Ratiner still believes that he and Pinto could have worked out a widely acceptable agreement on seabed mining (Ratiner 1989). But Paul Engo of Cameroon, the chair of Committee One, was angry that Pinto had operated behind his back and thus had shown contempt for the chair's knowledge of the subject and his negotiating skills. Toward the close of the Geneva session, therefore, Engo threw out the work that Ratiner and Pinto had done and published a pro-G77 Single Negotiating Text (SNT) that was totally unacceptable to the United States and other potential mining states.

At the session in New York in the spring of 1976, Engo shifted ground and accepted many of Ratiner's suggestions to make the new Revised Single Negotiating Text (RSNT) much more acceptable to the West. The central component of the RSNT was a parallel or dual track system of mining that provided equal access to the seabed both for public (that is, state-owned) and private entities and for the International Seabed Authority (ISA) through its mining arm, the Enterprise.

Engo's concessions that spring angered many G77 diplomats, who bypassed him by setting up small "workshops" (working groups) at the August-September session. Because Ratiner and other Western diplomats opposed the efforts of the G77 members of these workshops to reverse the concessions granted in the spring, the session ended in stalemate and acrimony. Thus, even as the Neptune Group became increasingly helpful to the negotiators in 1976, the conference itself appeared to be no closer to a widely accepted treaty in New York in September 1976 than it had been sixteen months earlier in Geneva. And the seemingly endless negotiations strengthened the hands of those in the U.S. government, media, and mining industry who wanted America to give up on the negotiating process and take unilateral steps on deep seabed mining.

Although the difficulties in the 1976 negotiations frustrated us, we were heartened by the Neptune Group's contributions. Following Sohn's advice to try to facilitate the negotiations by bringing people together to build trust and share ideas, we sponsored roughly fifteen seminars/brainstorming sessions from January through September. Several of these meetings focused on the most intractable problem facing the conference, seabed mining, while others dealt with such issues as dispute settlement and the delineation of the outer continental margin. In planning and scheduling these programs, we received invaluable help not only from the several capable student interns who joined us for each New York session, but also from other NGOs, including the Stanley Foundation, the Rockefeller Foundation, the American Friends Service Committee, the Quaker and Methodist UN programs, and Quaker House in New York (where many of the seminars took place).

In general, the programs we sponsored in 1976 and in subsequent years were better planned and more useful to the delegates than the ones we had put on in Geneva. Planning and usefulness improved for three main reasons. First, our emphasis in 1976 and thereafter was on problem solving by experts and by delegates rather than on propagating a common-heritage, pro-Third World ideology. Because extreme views in the Third World and in the industrialized nations had fostered polarization, we saw our task as strengthening the moderates in the conference, and believed that informational seminars and brainstorming sessions could help.

Second, instead of invariably following the standard format of a speaker or two followed by questions, several of our meetings lasted for two or three days and included ample time for informal brainstorming in small groups away from the pressures of the conference. Two basic types of programs emerged: the first, more traditional one involved experts as a resource for delegates and often began with talks; the second, more innovative one, Kimball recalled, "sometimes with and sometimes without outside experts, was to provide a forum for delegates holding opposite views to meet on third-party [nonofficial] ground in an informal setting and let their hair down, seeking compromise rather than having to espouse rigid national positions—the [Neptune Group's] honest broker role" (Kimball 1995).

Third and perhaps most important, we worked closely with delegates in setting the agendas for our meetings and then spent long hours on the phone trying to find speakers/brainstormers who, in Kimball's words, were "creative and good on their feet" and could "deal well with multicultural settings and people with very different levels of technical expertise" (Kimball 1995). In short, we did all we could to ensure that the programs we sponsored helped the negotiators rather than wasting their time.

One of our more useful programs, a seminar in Washington on January 30–31, 1976, led mainly by academic experts, examined the economics of seabed mining and the establishment and operation of the International Seabed Authority (ISA). The speaker whose ideas were most directly relevant to the conference was Prof. Louis T. Wells of Harvard Business School, an expert in negotiating land-based mining contracts for developing nations. After the meeting Sam sent the main ideas of Wells and other speakers to Paul Engo and to U.S. officials, who were working at the time to come up with proposals for the next session. A letter I wrote in June to Maxwell Stanley, the head of the Stanley Foundation, suggests how pleased Sam and I were about the meeting's impact: "You need to know that ideas generated at your session . . . went right into the decision-making process. The ideas were new and went to the Group of 77 and appeared in the

Revised Single Negotiating Text put forth by Chairman Paul B. Engo. In fact, they were also fed into the U.S. position and appeared as well in the April 9th speech of Henry Kissinger at the Hotel Pierre in New York" (Miriam Levering 1976a).

An even more useful program—at least for the Neptune Group's growing reputation as an information broker at the conference—was an off-the-record, brainstorming retreat at the Mohonk Mountain House in upstate New York on August 21 and 22. In this beautiful, relaxed resort setting, delegates and UN officials thrashed out such problems as financing the Enterprise, technology transfer, financial arrangements for seabed mining, and possible joint ventures with Dan Leipziger of the U.S. State Department, Joseph Hilmy of the World Bank, and Stephen Peterson of a law firm specializing in licensing agreements for technology transfer. In another room Barbara Weaver led a group wrestling with ways to prevent oil spills from ships and oil-spill pollution from drilling accidents. Her experts were C. J. Colman of Mobil Oil, Ken Demars of the U.S. Geological Survey, and Charles Koburger of the U.S. Coast Guard. Jim Bridgman, by now an experienced and effective member of the Neptune Group, commented on the retreat in his diary: "The weekend was a smashing success. It came at the right time when the negotiators wanted more time plus a change of scenery. Most of the key delegates came. We sprinkled in non-delegate experts and the sessions were very productive. The Mohonk session was so successful that for a while a 'Mohonk Group' met at the session to work out some details" (Bridgman diary n.d.).

The speakers, delegates, and UN officials generated several ideas that proved useful for the treaty. Equally important, those who attended grew closer to one another and found new sources of expertise to call upon when they were wrestling with difficult issues. All of us were grateful to Keith Smiley, the Quaker owner/manager of Mohonk Mountain House, for providing the free lodging and meals that made the retreat possible.

Not all of our seminars for delegates were so successful. A program at the spring session designed to get delegates from landlocked nations talking to delegates from coastal states flopped when the coastal-state delegates refused to attend. Similarly, delegates from communist China and Vietnam—nations then very much at odds—refused to attend a program in August on oil and territorial sea limits in the South China Sea. This seminar was not a complete failure, however:

it attracted delegates from the Philippines, Singapore, Indonesia, and Japan, and featured an excellent talk by Prof. Choon Ho Park of Harvard University, an expert on maritime boundary issues and conflicts in the region, followed by numerous questions and comments.

Besides getting delegates together with outside experts, members of the Neptune Group answered countless requests from delegates for information. One of the student interns, Jessica Mott, remembered providing "technical information" to delegates (Mott 1989). Bridgman recalled a specific example from the summer session:

> During the first week Ambassador Koh of Singapore indicated that I could be very useful to the session if I researched and obtained documents on the continental shelf and its oil potential. So I set to work in spare moments. A new edition of *Neptune* would be produced in two weeks and we decided to have a centerfold map showing fish potential, land area within and without an economic zone, actual oil production, potential oil production, present and future pollution potential, shipping. It became perhaps the best information source that we produced in *Neptune*. We all had to really dig for information. (Bridgman diary n.d.)

A situation that I found amusing developed during the summer 1976 session. Part of the fun of participating in the Neptune Group at the conference was sharing newly acquired information with other members of the group. Normally, I shared what I learned with other NGOs and with delegates, and much of what we picked up through conversations showed up in stories or in the "idea box" in *Neptune*. But one U.S. negotiator on seabed mining, whom we called "deep throat" after the main unidentified source in the Watergate scandal, insisted on confidentiality when he gave me inside information on the negotiations. I remember one occasion in which he confided to me in whispered tones in the garden room at Quaker House while other members of the Neptune Group played music and danced upstairs. Unfortunately, the student intern specializing in seabed mining resented my refusal to divulge the source of the information I was receiving in confidence.

As members of the Neptune Group developed closer relationships with negotiators and with UN officials, we were able to provide much more detailed information to journalists during the two New York sessions than we had at Geneva. Led by Arthur Paterson, we held briefings for the press before and after both sessions. Ably assisted by staff from Quaker House and from the American Friends Service Committee, we also put on six informational luncheons for journalists from the United States and abroad between March 2 and 23.

During the sessions, writers for the *New York Times* phoned repeatedly to get information on the conference, as did writers for the *Christian Science Monitor,* the *Washington Post,* and the *Philadelphia Inquirer.* We also were contacted by *Time, Newsweek, Business Week,* National Public Radio, the wire services, and other media outlets. In our press release at the close of the second session, we gave examples of concessions that we thought both the industrialized and the developing nations needed to make to complete the treaty.

During the Geneva session I had gotten to know Jim Brown, an editorial writer for the *New York Times* who also was a Quaker. Jim helped us line up journalists to attend our luncheons in March. He apparently did us an even bigger favor in September. On August 21 the *Times* published an editorial that, while balanced in some respects, put most of the blame for the deadlock at the conference on "third world zealots" and "extremists" (*New York Times* 1976a). The Neptune Group believed that the industrialized nations deserved their full share of the blame, and I (and perhaps others) let Jim Brown know how we felt. The *Times* editorial at the end of the conference was much more evenhanded in calling on both sides to move toward compromise (*New York Times* 1976b).

Whether or not we had influenced the *Times*'s shift in tone, Bernardo Zuleta, the Colombian diplomat who was the senior UN official for most of UNCLOS III, gave us full credit. As I was crossing the street one day at the UN, Zuleta stopped and thanked me for changing the *Times*'s editorial position. From then on Zuleta, who had been hostile to us and to NGOs generally in the past, welcomed our involvement in the conference.

Summing Up: Amazement, Anxiety, Determination

Three words—amazement, anxiety, and determination—best sum up how Sam and I felt at the close of the conference's fifth session in September 1976. We were amazed that a few lightly funded Quaker and Methodist activists, aided by six or eight recent college graduates and student interns, could have emerged in a little more than two years as by far the leading group of public interest NGOs at UNCLOS III. We were amazed and grateful that many—though certainly not all—delegates, UN officials, academic experts, NGOs, and journalists covering the conference appreciated our contributions and wanted us to keep working at future sessions. Edward L. Miles, a University of Washington professor with many friends at the conference, recalled that "both *Neptune* and the series of seminars became a very important, welcome addition to the enterprise. And by that time [1976], you had gained the trust of the [Group of] 77, and there was a lot of information flowing in both directions"(Miles 1990).

U.S. negotiator Leigh Ratiner commented that the seminars "tended to introduce a new sense of pragmatism into 77 thinking. What the Neptune Group did was that it helped in the acclimatization process—something that the negotiators [from the industrialized nations] could never do." From Ratiner's perspective the West had to have Third World support to complete the treaty; therefore, "the Neptune Group became extremely helpful to the negotiations, because you began to work directly on the Third World" (Ratiner 1989).

Ratiner was right: becoming focused during and after the session in Geneva in spring 1975 and gaining momentum thereafter, our work with delegates from developing nations constituted our most important contribution to UNCLOS III. Many delegates from small, newly independent nations wanted and needed our assistance in wending their way through the conference's legal, technical, and political mazes, and we were pleased to oblige. "We tried to help delegates with factual research and analysis, and we also helped them with the actual meaning of certain tortuously-drafted texts," Kimball recalled. "The former helped dispel suspicions; the latter was also helpful, but probably raised suspicions as well" (for example, about Western delegates' skill in inserting clever, often obfuscatory language into the negotiating texts that furthered their nations' interests) (Kimball 1995).

Desiring with all our hearts that the conference succeed, we were anxious in the fall of 1976 because we knew that it might well fail. As we have seen, a huge chasm separated the positions of developed and developing nations on deep seabed mining. And although we supported further U.S. concessions, we knew that additional concessions to developing nations, made in an atmosphere colored by demands for a new international economic order, might well mean that the U.S. Senate would not approve the treaty.

The depth of our concern about whether a treaty would be negotiated was reflected in a sentence filled with contingencies that Arthur Paterson and I included in the lead article of the October 1976 issue of *Soundings:* "If the occupant of the White House in 1977 can accommodate both legitimate third world concerns and U.S. interests, and avoid damaging unilateral action, participants give the Conference a 60% chance of success" (*Soundings* 1976).

Despite all of the anxieties and uncertainties, however, one thing was clear: we were determined to continue the fight for a treaty that we considered potentially the greatest step yet toward a world governed by law.

5

Helping to Shape the Treaty

THE FOUR YEARS FROM THE ELECTION of Jimmy Carter in November 1976 through his defeat for reelection in November 1980 were the most exciting and fulfilling ones in my life, in Sam's, and (I feel sure) in the lives of many others who worked with us on law of the sea. These years were exciting because the conference generally was moving toward broad agreement on a treaty that, while certainly not perfect from our standpoint, portended a major advance in world law and governance. They were fulfilling because, much more than in earlier years, we and our associates in the Neptune Group and in other NGOs were able to help formulate U.S. policy on law of the sea and, with even greater impact, to help the negotiations themselves move forward through what we saw as our catalyst or honest broker role at the conference. These years also were fulfilling because Sam and I never stopped being surprised and amazed that bright, idealistic young people kept showing up to work with us for little or no pay, and that leading officials in Washington and diplomats at the conference kept requesting our assistance and thanking us for our efforts.

The two leaders who contributed the most to our effectiveness during these years were the chief U.S. negotiator under Carter, Elliot Richardson, and the emerging star of the conference, Tommy Koh of Singapore. These two men not only liked us and developed close relationships with the Neptune Group; they also let many other officials and diplomats know that they considered us valuable contributors to the negotiating process. Because Richardson and Koh consistently encouraged our efforts, and because we knew that they shared our desire to achieve a treaty, our admiration of them, already substantial in 1977, grew greater with each passing year.

This chapter details the Neptune Group's largely positive experiences in participating in U.S. policy making and especially in facilitating negotiations from late 1976 through late 1980. But in our view these years also had a more negative side: the growth in America and in other large industrialized nations of increasingly vocal nationalist movements whose leaders opposed developing countries' efforts to use UN forums to enhance their influence in world affairs and disliked such "socialistic ideas" as stabilizing prices for raw materials through commodity agreements and viewing the minerals on the deep seabeds as the "common heritage of mankind."

The growing influence in the United States in the late 1970s of anti-UN, anti-Third World sentiment concerned us greatly, because it weakened support for the unavoidable compromises between developed and developing nations that were emerging at the conference. Our efforts to combat this nationalistic sentiment, including our continuing struggle against the passage of unilateral seabed mining legislation, provide the focus for chapter 6.

Assisting Pro-Treaty U.S. Officials

Jimmy Carter's election pleased Sam and me. We thought that the United States needed both new approaches that would be more sympathetic to the legitimate needs and concerns of the developing nations and a new negotiator on seabed mining whom other delegates might find more trustworthy and predictable than Leigh Ratiner had been. We also thought the U.S. policy making required more consistent involvement by the president, the national security adviser, and the secretary of state if it were to move away from the constant bickering among middle-level officials that had marked the previous years. Although U.S. policy did not receive more sustained top-level involvement under Carter than it had under Nixon and Ford, it did benefit from the first head of delegation with cabinet-level experience, Richardson, and from strong support for him from President Carter and, more directly, from Secretary of State Cyrus Vance. Richardson had held so many high-level positions under Nixon, in fact, that he was humorously introduced at the meeting of the Public Advisory Committee in April 1977 as "the *former* Elliot Richardson"![1]

1. The humorous reference to Richardson is in Sam Levering's notes on the meeting of the Public Advisory Committee on April 26–27, 1977 (Sam Levering 1977).

Rightly or wrongly, Sam and I thought in January 1977 that we had played a role in Richardson's selection. We thought so because we quickly became close to the member of Carter's transition team assigned to law of the sea, international law professor Samuel Bleicher of the University of Toledo. Bleicher invited Sam to bring his memorandum full of suggestions to his office, and then later visited our modest quarters in the FCNL building for long discussions of U.S. policy and the current state of the negotiations. Bleicher recalled these meetings and his role at the time as follows.

> I remember coming to your offices a couple times. Everybody else had an ax to grind—there was something they were trying to get ahead of.... It could well have been the first place that Richardson's name was mentioned. I remember the [first] meeting and feeling that the Leverings were doing important work, and that nobody else was taking the universal point of view that needed to be taken.
>
> I'm sure that his [Richardson's] name was on the list I gave to Anthony Lake, etc. But the decision was made by the Hamilton Jordan people, or by Cy Vance. I don't know how much influence my memos had. (Bleicher 1990)

Sam also worked closely with members of Congress and staff members sympathetic to our cause—notably Congressman Donald Fraser (Dem., Minn.), a leading member of the House International Relations Committee—to support the appointment of Richardson as part of a strong U.S. commitment to concluding the law of the sea treaty. When Carter appeared on Capitol Hill to discuss international relations in mid-January, Fraser asked him about his commitment to law of the sea. Carter responded that his friend Dean Rusk had apprised him of the importance of the negotiations, and that he agreed with the former secretary of state. The next day Fraser received a call from Stuart Eisenstadt, the head of Carter's foreign-policy transition team, requesting Fraser's detailed suggestions on law of the sea in writing. Fraser then phoned Sam for help on the memorandum.

Over the next few years Sam, Lee Kimball, our associate who was a full-time employee of the Methodist project during most of the Carter period, and I worked closely with Richardson and other officials in shaping U.S. policy. Richardson recalled the relaxed, informal nature of the relationship: "Miriam, or Sam, would come to see me in my office and tell me something that was going on, what they saw as a possible way of getting around a problem." Richardson believed that

Sam made a valuable contribution in the debates over U.S. policy in the Public Advisory Committee:

> He reflected a point of view that was aware of U.S. interests and which wished to avoid undercutting those interests, but he was also sympathetic and fair-minded to the Third World point of view.... He also, I think, kept steadily in view the importance of the treaty as a contribution to the extension of the rule of law, and thus to the achievement of a more peaceful world order. But he would come down on one side or another on an issue without any particular party line toward it. One had a feeling that he took each one on its own merits. And he was perfectly capable of speaking up on behalf of a mining industry interest when he thought that they had practical, realistic considerations in their favor. (Richardson 1990)

In the Public Advisory Committee meetings of the late 1970s, as in earlier ones, Sam urged patience in the negotiations and argued that U.S. seabed miners could only achieve secure legal rights for mining under an international treaty, not under unilateral domestic legislation. If that had been Sam's only contribution, however, his role in U.S. policy making would have been on the decline. For even the protreaty Richardson had concluded by late 1977 that the United States would have to move toward passing carefully drafted unilateral legislation to force militant Third World delegates to realize that seabed mining could go forward with or without a treaty. But Richardson expanded Sam's role in two ways: first, by permitting him to be part of the team that worked on U.S. policy on seabed mining issues beginning in late 1977; and second, by making him head of the Advisory Committee's subcommittee on the marine environment. Recognizing environmentalists' growing political clout and wanting them "to become an important constituency for the treaty," Richardson emphasized protection of the ocean environment much more than previous U.S. negotiators had done (Richardson 1990).

During the first several months of the new administration, Sam and I generally were pleased with its policies on law of the sea, which were more sympathetic to Third World concerns than had been the case during the Nixon-Ford years. We hoped that the conference would move quickly toward agreement, thus undercutting the mining companies' growing pressures on Congress to pass unilateral legislation. As we had hoped, the sixth session (May 23–July 15, 1977) was the most productive to date, with a negotiating group headed by

Norway's Jens Evensen making progress on several of the difficult issues relating to seabed mining. But then, as the session was about to end, a small group of militant G77 delegates got the ear of the Cameroons' Paul Engo, the head of Committee One, and induced him to remove the Evensen compromises and insert language ensuring Third World control over all seabed mining. Like the electrical outage that crippled New York in mid-July and gave us all stories to tell,[2] Engo's startling move cast a dark shadow on the conference and was talked about with amazement for years to come.

Engo's power play was a stunning blow to the negotiations, delaying the completion of the treaty and making it extremely difficult for Western negotiators to reverse Engo's changes sufficiently to obtain the kinds of provisions on seabed mining that the U.S. Congress and other Western legislatures would be likely to approve. Engo's action prompted a U.S. review of whether to continue with the negotiations, and contributed to the resignations of two key U.S. negotiators in Committee One, Richard Darman and John T. Smith. As Markus G. Schmidt has pointed out, the new chief negotiator for Committee One, career diplomat George Aldrich, "opened up the interagency [policy-making] process somewhat, ensuring that everyone could contribute to the drafting of position papers and giving them the lowest possible classification" (Schmidt 1989, 74).

Benefiting from this openness, Sam met frequently with Aldrich on seabed mining issues between late 1978 and 1980. Ably assisted by Lee Kimball, whose detailed knowledge of the post-1976 negotiating texts exceeded anyone else's in the Neptune Group, Sam sent lengthy memorandums responding to policy drafts that Aldrich sent from the State Department. Because Lee often was the principal, unacknowledged author of these memorandum, it was only fitting that Richardson named her to the Public Advisory Committee in early 1980.

Both Aldrich and Sam valued the relationship. Aldrich remembered Sam's "straightforward" approach to problems. In the work of the Public Advisory Committee, "Sam was always the plain-spoken,

---

2. The power outage, which began on the evening of July 13, caught me trying to get back to my apartment in midtown Manhattan. The failed stoplights snarled New York's traffic even more than usual, making it very difficult for me to cross streets, and caused Sam's train from Washington to stop in its tracks in suburban New Jersey. Other members of the Neptune Group were attending parties that continued by candlelight.

outspoken voice of reason" on seabed mining and environmental issues. "Elliot was fond of him, and I was too," Aldrich recalled. Disagreeing with mining representatives who tried to convince him that Sam was well-meaning but "impractical," Aldrich insisted that "Sam was the most practical person I could talk to." Aldrich also found Sam "helpful in battles with [Leigh] Ratiner and [John Norton] Moore," two former officials who now frequently criticized U.S. policy. Finally, Aldrich appreciated Sam's wide range of contacts at the conference. A "good source of information," Sam often had helpful suggestions about which delegates Aldrich should talk to (Aldrich 1990).

For his part, Sam was delighted to get to work with Aldrich on U.S. policy. "Aldrich would set up a subcommittee in the State Department," Sam recalled. "Most of the others were too busy to show up, but I would be there. Aldrich liked to have me there because he knew I would be there to help" (Sam Levering 1990).

Sam often did "help"—for example, in Aldrich's efforts to simplify the treaty's seabed mining provisions in the winter of 1978–79 and in the State Department's fight with the Treasury Department over State's insistence that mining companies be allowed to take credits to reduce U.S. taxes for some of their payments to the Authority for the right to mine the seabed (Public Advisory Committee 1978, 116). But he also opposed Aldrich and Richardson at times, notably on the U.S. concession at the spring 1978 session that accepted the mandatory transfer of Western technology to ensure that the Enterprise would be able to conduct mining operations. Aware that U.S. mining representatives and leading members of Congress adamantly opposed any requirement to transfer technology, Sam criticized mandatory transfer at the Public Advisory Committee meeting on March 8, 1979, arguing that "the approaches that are being taken are really unfortunate" (Public Advisory Committee 1979a, 65–67).

Although he sympathized with Sam's concerns, Aldrich responded that it would be "impossible in the Conference to do away with [Third World] suspicions which will require some kind of obligation of technology sale" (Public Advisory Committee 1979a, 68–69). Regrettably, both men were correct: Aldrich was right that the G77 would not agree to a treaty that omitted the mandatory transfer of seabed mining technology; and, as the next chapter demonstrates, Sam was right that mining companies and Congress would oppose any treaty that included it.

While working with Leigh Ratiner, Marne Dubs, and other representatives of the mining companies who also served on Aldrich's committees, Sam became convinced that these men were making valid points about the industry's needs and the treaty's deficiencies. Indeed, Sam could not visualize a workable treaty that did not encourage industry to undertake deep seabed mining. Given the Neptune Group's known sympathy for developing nations, Sam's concern for the needs of the prospective miners surprised Richardson, who once told him so. When asked later about his views, Sam commented: "I guess I have always been a realist. I think business experience makes you a realist" (Sam Levering 1990).

Although all members of the Neptune Group agreed with Sam about the need for a workable treaty, the rest of us showed greater sympathy for the G77 on seabed mining issues—and less concern for the industry—than Sam did. Having worked closely with the industry and Congress in the late 1970s, Sam was less surprised and shocked than other members of the group were when the new Reagan administration turned sharply against the negotiations in 1981. "Dubs said we didn't improve the treaty" sufficiently from the miners' standpoint, Sam recalled. "If so, it wasn't because we didn't try" (Sam Levering 1990).

Supporting the Efforts of Environmentalists

On seabed mining, Sam and Lee Kimball worked long and hard with Aldrich, Richardson, and other officials in shaping U.S. policy. Sam, Lee, and others in the Neptune Group also focused in these years on blocking unilateral legislation in Congress. Sam's and my interest in environmental protection, though sincere, was definitely secondary—partly because we were part of the Woodrow Wilson–Franklin Roosevelt generation of world order activists, not the more recent generation of environmental activists. Nevertheless, Sam and others in our group helped make it possible for the small but growing number of environmental activists on ocean issues to work with Richardson and other U.S. officials and foreign diplomats to make changes in U.S. policy and in the negotiating text during the late 1970s.

Richardson deserves much of the credit for the environmental provisions that worked their way into the treaty during his tenure. An avid bird-watcher, Richardson not only shared the environmentalists' goals, but also encouraged and supported them both in Washington and at the conference. Often, for example, he overruled subordinates' reservations about pushing too hard on environmental issues. At the conference he also fought for tougher environmental provisions. One example: sporting a whale tie, he accompanied a leading activist, Patricia Forkan of the Humane Society, to a key meeting with Soviet diplomats in 1979 to argue for stronger protection for cetaceans.

Most of the remaining credit belongs to a small group of intelligent, determined American environmentalists, notably Forkan, Anita Yurchyshyn of the Sierra Club, Jim Barnes of the Center for Law and Social Policy, and Robbins Barstow of the Connecticut Cetacean Society. Alerted by his nephew, Arthur Paterson, to the inadequacies of the current negotiating article on whales (article 65), Barstow publicized the plight of whales by bringing a petition with twenty-two thousand signatures to the 1977 session in New York. Barstow's group declared Sunday, May 22, 1977, to be "International Whale Day," held a rally across from the United Nations cosponsored by twenty conservation organizations, and distributed to delegates an eight-page booklet, "Whales and the Law of the Sea."[3]

Again with help from the Neptune Group, Barstow stirred up environmentalists in Washington on the issue and gained the support of Sen. Lowell Weicker (Rep., Conn.) in pressuring the U.S. delegation to work to protect cetaceans. When article 65 was strengthened with U.S. support, the ever-exuberant Barstow sent thank-you notes to the negotiators on behalf of the whales. In this case, the Neptune Group—notably Paterson—served as a catalyst for fellow NGO activists.

Richardson put Barnes, Forkan, and Yurchyshyn on the Advisory Committee and, indeed, on the U.S. delegation itself. Along with others including Anthony Wayne Smith of the National Parks and Conservation Association, these environmentalists served on the Marine Environment Subcommittee, where they successfully prodded the chairman, Sam, to send detailed memos (often written by Barnes) to Richardson and other U.S. officials and set up meetings with them to deal with unresolved environmental issues.[4]

---

3. Arthur Paterson recalled that "fewer than one-hundred people, perhaps fewer than fifty" attended the rally Barstow organized. But "this poor attendance did not deter him" (Paterson 1995).

4. For example, an environmental working group, including Sam, met on July 11, 1978. Sam followed up the next day with a letter to Elliot Richardson requesting that he meet with the environmentalists in August.

What did the environmentalists accomplish? Perhaps their greatest achievements were in elaborating the requirements for environmental regulations for deep seabed mining and in strengthening article 65 of the treaty, which sought to protect marine mammals. In Washington, Barnes and Yurchyshyn provided leadership in ensuring that the U.S. seabed mining legislation incorporated numerous provisions to protect the marine environment. At the conference, Yurchyshyn and Forkan fought effectively for protection of the marine environment in general and preservation of marine mammals in particular.

Besides Richardson's support, the keys to these environmentalists' success were thorough knowledge of the negotiating texts and dogged determination to succeed. Yurchyshyn recalled an encounter with an unfriendly American official that occurred on the day she became a member of the U.S. delegation: "He asks me out to lunch, my first day. I thought, what a nice, collegial thing to do. And then, without any niceties or small talk, he leaned across the table from me and said, 'Look, honey, you don't expect the United States government to spend its negotiating capital on your paltry environmental amendments, do you?' " (Yurchyshyn 1990).

Judging correctly that she had Richardson's support, Yurchyshyn did not let this hostile comment intimidate her. Nor did she lose heart when Soviet delegates refused for more than two sessions to support a provision requiring the environmental monitoring of seabed mining. She recalled what happened after the Soviets agreed, at the last minute, to accept it: "We finally resolved our differences [with the Russians], and then had to get it once again onto Engo's desk. And the door was closed. He said, 'It's too late, it's too late, go away.' And I remembered knocking and I wouldn't go away, and I said, 'Chairman Engo, this is Anita. I've got good news.' And he said, 'It's too late.' And I kept knocking and saying, 'But you *must* listen.' And he finally got it in" (Yurchyshyn 1990).

In a meeting of the Public Advisory Committee on May 18, 1979, Richardson cited the environmentalists' contributions at the most recent session in Geneva as "an outstanding example of the kind of help that we have received so consistently from members of this Committee" (Public Advisory Committee 1979b, 67). Aldrich's more detailed comments at the same meeting provide one of the best summaries of what NGOs, at their best, can accomplish in international negotiations:

[T]he progress we made on the environmental protection of seabeds this time is owed entirely to the initiative taken by the members of the Advisory Committee who attended, representing the Environmental Subcommittee, and not simply in taking the initiative to force us to look at these problems, but in talking to the Secretariat, in lobbying other delegations and, in a very real way, in gently twisting people's arms, particularly at the very end of the session. We found that it took a lot of time and effort by Anita Yurchyshyn, who went around to see Paul Engo and the Soviets and everybody else, and I think it is a good example of what we can do in a cooperative way. (Public Advisory Committee 1979b, 67)

Finally, what, besides Paterson's dealings with Barstow on whales, did the Neptune Group contribute to the environmentalists' work? According to Richardson, "Sam played a very valuable role in helping to identify what their interests were, and to get them to focus on what were achievable ways of meeting these interests. The problem with the environmental people, in the beginning particularly, was that they didn't have a real feel for the Third World attitudes" (Richardson 1990). Compared with some environmentalists, in other words, Sam had greater experience and sensitivity in working with both U.S. and foreign negotiators. One U.S. official, George Taft, remembered the "pressure" from environmentalists (Taft 1990); another, Bernard Oxman, criticized their "very young, very bright, extremely aggressive lawyers" (Oxman 1990). Taft praised Sam for being "levelheaded," and Oxman observed that "he had a far more mature sense of how to talk to the people across the table" than did some of the environmentalists.

Yurchyshyn believed it helped that Sam was "extremely respected" on the Advisory Committee (Yurchyshyn 1990). And the fact that delegates liked and admired the Neptune Group aided environmentalists at the conference as well. When Patricia Forkan arrived in Geneva in April 1979, for example, members of the U.S. delegation advised her to work with the Neptune Group because they were respected.

Five Ways We Helped the Negotiations

Although members of the Neptune Group contributed to U.S. policy on the deep seabed and the marine environment, our greatest contributions during the Carter years came in our efforts to assist the negotiations. Our contributions occurred primarily in five areas, most of which were continuations of the kinds of work we had done at the conference between 1974 and 1976. The key difference was that, building on experience, we were more effective in every area than we had been earlier.

The first area in which we contributed was getting delegates together informally in parties and receptions. Typically organized by Barbara Weaver of UMLSP and other members of the Neptune Group, including Sister Mary Beth Riessen and me, these usually were held at the Methodist Center at the United Nations or at Quaker House in Geneva. These gatherings became ever more popular as the Neptune Group's status in the conference rose: in 1978, for example, Richardson and other key U.S. delegates began to come fairly regularly.

One explanation for these receptions' popularity was that they were much less formal—and hence less stuffy—than the formal receptions that delegations organized. Another explanation was that the mostly male delegates—many of whom were without regular female companionship during conference sessions—enjoyed talking with the women of the Neptune Group, at least two of whose regular members (Barbara and Lee) were young, good-looking, and unmarried. Arthur Paterson noted the "charm of the young women" that gave them a "better entrée to the diplomats" than he and Jim Bridgman had. As for me, Paterson "remember[ed] Lee saying that somewhere in the mind of some of the diplomats . . . your constantly being there, your stature, your Quaker ways, and your age all contributed to a sense of deference and respect" (Paterson 1989).

A second area of contribution was setting up press conferences and media briefings, normally before and after each session and often featuring Tommy Koh—a favorite with journalists because of his straightforward, honest approach. During these years we received many compliments from journalists from major newspapers, magazines, and television networks who said that these meetings—plus numerous informal conversations with us—helped keep them abreast of developments in the negotiations. At times portions of our press releases on trends in the conference were paraphrased in major news stories, as in an August 1978 Associated Press article on seabed mining legislation.[5]

5. A sentence from a Neptune Group press release of August 17, 1978, modified slightly, appeared as the second sentence in an Associated Press article (Koo 1978).

Third, we published one issue of *Neptune* during six of the seven conference sessions from 1977 through 1980, and two issues at the Geneva session in the spring of 1978. Compared with the nine issues published in 1975 and 1976, these eight issues gave relatively more attention to possible solutions to the remaining difficult problems facing the conference, and relatively less to whether developing and geographically disadvantaged states were being treated fairly in the negotiations. Given the substantial distance separating developed and developing nations on seabed mining and related issues, we decided that *Neptune*'s focus should be on finding solutions satisfactory to both sides, not on championing the interests of one side over the other. As in the past, *Neptune* frequently featured ideas generated in the many seminars and brainstorming sessions that the group sponsored. During the late 1970s and early 1980s, in short, *Neptune* contributed to the ongoing discussion of unresolved issues and enhanced the group's reputation as an honest broker at the conference.

During these years delegates, journalists, mining company representatives, and other NGOs frequently stopped us, usually to praise *Neptune*[6] but also at times to ask if we could include their articles or ideas in it. Delegates also asked why we did not publish multiple issues per session, as we had in two of the three sessions in 1975–76. We published relatively few issues (usually one at the beginning of each session) partly because we had limited funds, but even more because we believed that our other activities were now more useful to the conference.

Fourth, we were involved in a wide range of unpublicized efforts to facilitate the negotiations. For example, several members of the Neptune Group, including Lee Kimball and me, used our many contacts both inside and outside the conference to answer countless factual questions from delegates. This was an important service, because delegates from developing countries would have been too embarrassed to ask for information from delegates from developed countries or from representatives of mining companies. Because Lee

---

6. Among the positive comments about *Neptune* were ones offered by journalists near the end of the spring 1978 session in Geneva. Jon Rollow of the *Washington Post:* "Very useful check on all the rhetoric. One of the few objective sources of information." Alan McGregor of the *Times* (London): "This publication is absolutely essential to help so many delegates to the law of the sea conference understand what it's all about—to say nothing of the media." These comments are taken from Diamante 1978.

was one of the handful of true experts on the seabed mining negotiations by the late 1970s, she often could answer delegates' questions herself, and I and others in the Neptune Group often checked with Lee before getting help from other experts. Besides her articles in *Neptune* and other publications, Lee also prepared options papers for the negotiators' use and served as an informal mediator between them on possible compromise language. In short, much of the Neptune Group's most valuable work at the conference was done in confidential discussions with negotiators and in research and brainstorming for them, all designed to move the negotiations forward as rapidly as possible.

Finally, we sponsored numerous seminars, retreats, and brainstorming sessions—at least forty altogether from 1977 through 1980—that became our best-known and most-appreciated contribution to the conference.[7] These ranged from luncheon meetings with one speaker followed by questions to weekend retreats with several speakers and ample time for both formal discussions and informal conversations. Most of these meetings took place while the conference was in session, but other important ones—such as the one at Bryn Mawr College in November 1977 on the definition of the outer continental margin—took place between sessions.

The twofold, closely linked purpose of many of these meetings was (1) to allow delegates to meet informally with outside experts to discuss ideas that might lead to breakthroughs in the negotiations and (2) to set up situations away from the conference where the different sides in disputes among the delegates could stop posturing and begin to find a solid basis for compromise (Kimball 1995). Because leading negotiators appreciated these meetings and kept asking us to organize more, and because we repeatedly saw ideas that emerged from the discussions appearing in the negotiating text, we too believed that these seminars and brainstorming sessions were the Neptune Group's greatest contribution to the conference.

The focus of most of our seminars and brainstorming sessions involved problems associated with setting up the seabed mining system. As the other major issues facing the conference had already been resolved or were well along the way to being resolved, the seabed mining regime engaged many of the best minds both within and

---

7. For scholarly references to—and praise for—Neptune Group programs at the conference, see, for example, Raiffa 1982, 283; Sebenius 1984, 31; and Sanger 1987, 33.

outside the conference throughout the Carter years. Much progress was made, albeit slowly. Regrettably, however, the issues surrounding seabed mining were not entirely settled in the seven meetings of the conference from 1977 through 1980.

Assisting the Negotiations, 1977–1978

Our first brainstorming session of the Carter years, which took place at the Carnegie International Center in New York from February 4–6, 1977, set the pattern for our work during this period. The focus was the system of exploitation of the deep seabed—a very broad subject, as it turned out. Among the many ideas discussed at these meetings, at the time two seemed especially promising: first, in the interest of getting a treaty it made sense to work on trying to make Kissinger's proposed parallel system (Enterprise plus private or state-owned mining contractors) workable; and second, it also made sense to set up a provisional mining system (that is, one lasting perhaps twenty years) that then could be reviewed and modified at the end of a specified period. To no one's surprise, the discussions reflected deep divisions between the views of the three U.S. officials and those of the four representatives of Third World nations who attended: Antonio Gonzales de Leon of Mexico, Tommy Koh of Singapore, Samir Sobhy of Egypt, and Christopher Thomas of Trinidad and Tobago. At this conference even Koh, whom we considered a moderate, frequently sounded like a spokesman for G77 views ("A Report . . . " 1977).

This brainstorming clearly was useful as a place where the United States and G77 representatives could exchange ideas and get to know each other better. But it turned out to be even more important that these officials got to hear the views of three other participants: Joseph Hilmy of the World Bank, Prof J. D. Nyhart of M.I.T., and David N. Smith of the Harvard Law School. These outsiders to the negotiations offered fresh suggestions on such issues as how an interim system of exploitation could be set up and how mining companies might enter into contracts with the Authority.[8] Thereafter, Koh and

---

8. In the seminars in which they participated, David Smith and Louis Wells drew upon some of the ideas from their 1975 book *Negotiating Third World Mineral Agreements*. We believed that the experiences of Third World governments in negotiating mining contracts with multinational corporations could contribute to the drafting of fair, practical seabed mining provisions in the treaty.

other conference leaders repeatedly asked us to arrange meetings with outside experts, most frequently with ones (such as Nyhart and Smith) affiliated with M.I.T. or Harvard.

Our efforts to assist the negotiations by encouraging moderation and realism among G77 delegates continued into the spring and summer. Lee Kimball organized a seminar on April 17 that sought to make the Enterprise concept more workable through realistic analysis of its financing and business structure. The leaders were Charles Houseman of the Chase Manhattan Bank and Walter Chudson of the UN Center for Transnational Corporations. After the New York session of the conference began on May 23, Koh and other delegates urged us to bring in additional experts on mining issues. We obliged by asking Prof. Louis Wells of Harvard's Business School to spend June 14 in New York with representatives of the UN secretariat carefully going over U.S. and Indian proposals on financial arrangements between seabed mining companies and the Authority and meeting with Third World delegates. A letter I wrote at the time commented on Wells's contribution:

> We had Louis Wells down to go over the fine print of two proposals on financial arrangements: that of the U.S. which he found full of holes where the companies could take advantage if they remained, and even more so that of India. He helped [Dutch diplomat Hans] Sondaal and [Australian diplomat John] Bailey for two hours with this and then had dinner with four key developing world delegates including the author of the Indian proposal. With great finesse, he pulled them along quite a ways. (Miriam Levering 1977a)

At a weekend retreat at Lake Mohonk on June 24–26, David Smith (who had been so helpful at the February brainstorming) and Thomas Waelde of the UN Center for Transnational Corporations offered useful perspectives on financial arrangements. Like Wells, Smith had long and extensive experience in helping developing countries negotiate mining contracts. Afterward, several delegates told us how valuable this weekend retreat had been. Another expert on negotiations between developing countries and mining companies, New York lawyer Charles Lipton, spoke at one of our six luncheon seminars during the session.

Our emphasis on financial arrangements at this and subsequent sessions stemmed partly from the realization that no treaty would ever be signed that did not clearly spell out financial terms for seabed mining operators and funding for the international bodies that the treaty established. The delegates from the industrialized nations had to know how the treaty would affect their private mining companies and public treasuries to be able to justify these arrangements before their national legislatures.

Two developments at the beginning of the session in May highlighted different components of the Neptune Group's work. The intellectual component was apparent in the twelve-page issue of *Neptune* that we distributed to delegates, journalists, and NGOs at the beginning of the session. Roughly half of the articles—including Lee Kimball's two-page discussion of how to structure the Council of the International Seabed Authority (ISA)—were devoted to seabed mining. Providing relevant data on which decisions could be made, Lee laid out three possible ways of setting up the voting system in the Council. We subsequently learned that the Evensen Group—the main negotiators on Committee One issues at the 1977 session—had used Lee's article in their deliberations. On a lighter note, *Neptune*'s back page featured a "Mal de Mer [seasickness] Game" that allowed delegates to laugh about the difficulties the conference faced along the road to a treaty. As usual, the many compliments that came our way provided more than enough compensation for the hard work and long hours every issue required.

The Neptune Group's spiritual side was expressed in an ecumenical prayer service at the church center at noon on May 23, the day the session began. Organized by Barbara Weaver and involving Methodist law of the sea activists from across America as well as participants from religious groups in New York City, the prayer service was intended "to speed progress in the deadlocked Law of the Sea Conference" (Invitation 1977).

The well-attended service was followed by a procession to the nearby "Isaiah Wall," so named to remind us of the Jewish prophet's admonition to beat swords into plowshares. At the outdoor ceremony Jim Bridgman released about one hundred gas-filled balloons. Passers by were startled as they floated skyward over the United Nations, carrying our hopes and prayers for UNCLOS III with them.

Our prayers seemed even more needed at the end of the session in mid-July, when Engo's rewriting of the Committee One negotiating text from a G77 perspective raised the possibility that the United States and other industrialized nations would withdraw from the negotiations. Fearing that possibility, we spent much of the remainder

Jim Bridgman and Sister Mary Beth Reissen with other leaders of the ecumenical prayer service in New York, 1977. Courtesy of Miriam Levering.

of 1977 urging U.S. officials and G77 moderates to view Engo's move as only a temporary setback. We also hoped that Engo's action would lessen his influence and that of other G77 hard-liners in the conference, thus strengthening the hands of moderates such as Koh.

That hope turned out to be justified, for new rules adopted at the spring 1978 session limited the ability of committee chairs to make unilateral changes in the text.

The pivotal year for the negotiations—and for the Neptune Group's ability to help them move forward—was 1978. At the Geneva session in the spring, seven negotiating groups were set up to tackle "hard core" issues that were preventing the completion of the treaty. Three of these negotiating groups (NG 1, NG 2, and NG 3) focused on Committee One issues. Frank Njenga of Kenya, the youngest and one of the smartest negotiators, headed NG 1, which dealt with issues relating to access to the deep seabed. This negotiating group was given a variety of thorny problems, including whether the awarding of contracts should be automatic or discretionary, whether limits should be placed on seabed mining to protect land-based producers, whether technology transfer should be mandatory, how long mining contracts would last, and how many contracts would be available to one operator or even to one nation as seabed production was permitted to expand relative to land-based production.

NG 2, headed by Tommy Koh, focused on financial arrangements, which included all the politically sensitive and often highly technical issues relating to who would have to pay and how much they would have to pay. Koh's group would have to set up payments systems and amounts that would satisfy miners and fund the Enterprise, and they would have to do so in an area (the economics of prospective seabed mining) in which hard information was scarce indeed.

NG 3, focusing on the structure and functions of the Authority, was headed by Paul Engo. This difficult issue was likely to be among the last to be decided, and decided largely among the larger countries—and among blocs of smaller countries—to ensure that no important nation or group of nations could be overrun by others. In fact, Koh ended up being the key negotiator in this area, an arrangement Engo accepted so long as Koh was careful to give the credit to Engo.[9] These were the three most important negotiating groups,

---

9. Koh recalled his relationship with Engo as follows: "Because we got along well and I knew how to manage him—which is basically to do the work for him but let him take the credit—he began to give me more and more work to do. He asked me to help him on the Council [of the ISA], and on the decision-making procedure" (Koh 1990).

because everyone knew by 1978 that the negotiations would succeed if the major Committee One issues were resolved.

Because the leaders of the Neptune Group were close personally to Koh, and because this leader of the conference wanted our help in moving the negotiations forward, we were able to make a greater contribution in 1978 than we ever had before. After Koh became head of NG 2, he increased his requests for seminars/brainstormings on specific Committee One issues. But it was not just Koh: in late 1977, for example, both Elliot Richardson and Bernardo Zuleta urged us to step up our private gatherings of delegates and experts.

Responding to these requests, we worked with the Stanley Foundation in organizing a brainstorming conference on financial arrangements in New York on January 28–29, 1978. The participants included Zuleta; four G77 diplomats, including Koh; eight officials, business executives, and academics from the United States and western Europe; two representatives from the Stanley Foundation; and Sam and me.

The big hit of this gathering was Prof. J. D. Nyhart of MIT, who introduced delegates to the MIT computer study he directed on the

Three leaders of UNCLOS III: Bernardo Zuleta, Hamilton Amerasinghe, and David Hall. Courtesy of the United Nations, photo by Y. Nagata.

profitability to seabed mining companies of operations from each mine site under either a three-metal or a four-metal recovery operation. Nyhart presented figures on expected expenditures and income, and hence expected profits, under various scenarios. The U.S. Department of Commerce had funded this study over several years, and it had become Lance Antrim's master's thesis under Nyhart's direction.

At the time of this brainstorming, Koh and other delegates were looking for ways to introduce credible figures into the negotiations on financial arrangements, and quickly saw the potential of the MIT study. Koh met Nyhart at this seminar and arranged with him to visit Cambridge, where he could learn the details of the study and quiz Antrim and his associates to make a fuller assessment of the study's strengths and weaknesses. One of Antrim's associates on the study, Harvard graduate student James Sebenius, attended the brainstorming and helped Nyhart explain the MIT study.[10] Just after this meeting, Zuleta wrote to thank the members of the Neptune Group for helping all involved in the negotiations achieve "a greater understanding of the crucial issues that will have to be resolved in order to make a comprehensive Law of the Sea treaty possible" (Zuleta 1978).

Several members of the Neptune Group, including Lee Kimball and me, went to Geneva for the seventh session of the conference, which met from March 28 through May 19. Before the session, Richardson had warned that the United States would proceed with unilateral seabed mining legislation if the conference did not move quickly toward consensus, but this warning had little effect. Instead of negotiating, the delegates spent the first three weeks of the session discussing whether Hamilton Amerasinghe of Sri Lanka could continue to serve as president of the conference even though a new government back home had removed him from the delegation. When this matter was finally decided in the affirmative, the seven negotiating groups began serious work. But the continuing divisions between developed and developing nations prevented major breakthroughs on Committee One issues.

Between April 1 and May 2, the Neptune Group sponsored three luncheon seminars and four evening panel discussions. Although most of these meetings featured top negotiators, they were more informational and educational than problem solving, and hence were not as

---

10. For a summary of the discussions at this conference, see "A Report . . ." 1978.

Sam Levering (left) with European NGO representative Adolf Schneider at Geneva, 1978. Courtesy of Lee Kimball.

helpful to the negotiations as the brainstorming sessions we had held in New York in February and at the conference in June and July of 1977. At the panel discussion on April 11 that focused on seabed mining issues, Alvero de Soto of Peru, a spokesman for the G77 on Committee One issues, argued that the International Seabed Authority should be a "model" for the new international economic order. "If we try to load all of the ideology onto this one conference, it may collapse under all this weight," U.S. negotiator Ronald Katz responded. The third panelist, Tommy Koh, came across as a moderate by contending that acceptable compromises were possible between developed and developing nations ("Summary..." 1978, 4–11).

At the end of April I suddenly became sick and was diagnosed with very high blood pressure. Sam was in Geneva with me and, because we could not afford the large deposit required for admission to a Swiss hospital, we decided to return to America and seek medical treatment for me there. Drugs brought the high blood pressure under control, and my doctor advised me to stop putting in the work weeks of seventy hours or more that had been common during the past several years. Although this illness and the ongoing need to take several pills a day were unwelcome reminders of my age (sixty-four) and my mortality, I was determined to continue to work nearly as hard as ever.

Lee Kimball and nine other experienced and novice NGOs associated with the Neptune Group did a good job in Geneva during the remaining three weeks of the session, getting an issue of *Neptune* to the printers on May 9 and, in Lee's case at least, keeping close tabs on the negotiations. "One problem with Final Clauses and [Frank] Njenga's text is proposal for liberation groups to be party to treaty [in Final Clauses] and in Njenga that 'peoples' be taken into account," Lee wrote Sam and me on May 10. "I've spoken with Njenga contact on this and the red flag reaction in Congress and something may be able to be done—trouble is, it's already in the works" (Kimball 1978). On this issue and on others such as mandatory technology transfer, it is an understatement to note that U.S. and other Western diplomats were not dominating the negotiations in Geneva that spring.

The Neptune Group made its largest single contribution to the conference at the resumed seventh session, which met in New York from August 21 through September 15, 1978. Koh had become convinced over the past six months that the MIT computerized econometrics model offered the best chance to break the deadlock between developed and developing nations over financial arrangements for seabed mining. He therefore asked the Neptune Group to organize an all-day seminar on the MIT study on Saturday, August 29, and he personally called around to make sure that a large contingent from the G77 were among the roughly fifty people (including thirty-six delegates and three mining company representatives) who attended the meeting at the church center across from the UN. With financial help from the World Affairs Council of Philadelphia, we were able to bring the entire MIT team of J. D. Nyhart, Lance Antrim, and Dale Leshaw of MIT, plus James Sebenius of Harvard.

During one three-hour session in the morning and another in the afternoon, those in attendance listened to a thorough explanation of the model by Nyhart and the other members of his team, and then got to observe how well the MIT/Harvard team held up under a lengthy barrage of questions from critics ranging from free-market industry representatives to G77 supporters of state-run enterprises. Koh did an excellent job as chair of the session. The Cambridge academics also handled themselves admirably: they maintained a stance of scientific objectivity while making it clear that the MIT model was not a certain set of facts but reflected their best estimates based on current knowledge of the projected economic conditions facing the seabed mining industry. Because the background of most delegates was in law rather than in economics, much of the day was spent explaining such economic concepts as "cost model," "accounting model," "net present value," and "internal rate of return" ("Seminar..." 1978). Sebenius later offered a perceptive summary of what had happened at the seminar and how the members of the MIT team had established their impartiality:

> Delegates of all political persuasions packed the politically timely and visibly Koh-blessed "MIT Seminar," which featured the principal members of the MIT team. Over the course of the day, they explained their model and discussed factors affecting future seabed profitability.
>
> Listeners vigorously questioned many of the model's assumptions, and, in particular, its "baseline" values. The team's usual response to queries and challenges was to explain the source of the questioned assumption and to demonstrate the sensitivity of the model's results to the factor in question. That technique highlighted the underlying technical and economic uncertainties, but it also seemed to enhance the credibility of the effort. (Sebenius 1981, 84)

The modesty and effectiveness with which the MIT/Harvard team presented their model was a turning point in the negotiations. For the first time, there was a common base of knowledge and assumptions upon which those participating in Koh's negotiating group could work toward consensus. Besides its thoroughness, U.S. negotiator Ronald Katz noted another advantage of the model: because it was an academic study, negotiators could "accept the MIT model as a basis for negotiations in a face-saving way without conceding the correctness of the point of view of their opponents" (Katz 1979, 210).

Not everyone was enthusiastic: mining company representatives thought the projected profits were too high (Dubs 1990), and Paul Engo complained that the delegates had "been dragged into adopting models and systems of calculations on fictitious data that no one, expert or magician, can make the basis of any rational determination" (Sebenius 1981, 85). But Engo and other skeptical G77 delegates went along with Koh, partly because mining company representatives and officials from Western nations were criticizing the model and partly because there seemed to be no other way to move toward consensus on financial arrangements. Indeed, the only major advance in the negotiations at this session was a paper Koh produced, based largely on the MIT model, that specified figures and percentages for payments from private contractors to the Authority.

Our sense of elation that we had contributed to the negotiations at the resumed session was kept in check by two realities: first, there were still many details to work out before consensus could be reached on a treaty; and second, Elliot Richardson favored the passage of seabed mining legislation that fall to show his displeasure with perceived G77 intransigence and thus, he hoped, to move the negotiations forward. In a statement at the end of the resumed session, Richardson said he felt both "encouragement" and "dismay," but he was "still hopeful that impasses can be broken and a treaty struck" ("Statement . . . " 1978).

Publicly, Sam and I were as upbeat as usual; but privately, during one meeting I released some of my frustration at the seemingly endless negotiations by jotting a few lines from Lewis Carroll's *Through the Looking-Glass* in which the Walrus and the Carpenter were discussing how "grand" the seashore would be if "this [the sand] were only cleared away":

> "If seven maids with seven mops
> Swept it for half a year,
> Do you suppose," the Walrus said,
> "That they could get it clear?"

A very gratifying moment occurred at an all-day brainstorming on November 4, 1978 on decision making in the International Seabed Authority. At the end of the meeting Bernardo Zuleta rose to thank the members of the Neptune Group for the help we were giving to the conference. He said he also wanted to atone for remarks that Lee Kimball had overheard in Geneva: that nongovernmental organizations were in the way and were backseat drivers, always calling for impossible things in this un-ideal world. He commented that we were the kind of backseat driver who watched the map and instead of berating the driver for getting sleepy, announced that the next rest area was three miles ahead and said, "Let me buy you a cup of coffee."

Coming from a high UN official who had been so critical of us, this praise meant much to all of us. It reinforced something we had known without doubt after the seminar on financial arrangements in

August: the Neptune Group had truly come of age as a participant in the negotiations.

Continuing Our Facilitating Role, 1979–1980

As in 1978, in 1979 the conference met in Geneva in the spring (March 19–April 27) and in a resumed session in New York in the summer (July 19–August 26). Some progress was made on seabed mining issues in Geneva, especially in the last few days, leading Richardson to state that the newly revised negotiating text offered "a substantially improved prospect of consensus" (*New York Times* 1979, 20). Enough additional progress occurred in New York for Richardson to assert that "the successful completion of the Third U.N. Conference on the Law of the Sea is now in sight."

Despite the progress and Richardson's public optimism, many details remained to be resolved in subsequent sessions. Moreover, as the next chapter documents, mining representatives and members of Congress were increasingly vocal in opposing several of the emerging compromises on Committee One issues.

From our standpoint, 1979 was another highly successful year in the Neptune Group's effort to facilitate the negotiations. We again were proud of the numerous seminars and brainstorming sessions that we sponsored, which were better attended than ever before.[11] Our two issues of *Neptune* were well received, and we continued to get many requests for information, especially from G77 delegates. During the Geneva session, for example, a G77 delegate handed Sam and me a paper at nine one morning requesting that we bring to his seat in Salle XX by five o'clock the latest information on seabed mining. We worked frantically throughout the day, using both published materials and phone conversations to put together the most objective information we could find from the several countries with potential seabed mining firms. By five we had a concise summary on his desk, which we then expanded into an article for the upcoming issue of *Neptune*.[12]

---

11. We sponsored six seminars and brainstorming sessions between March and May, mostly in Geneva, and another seven in New York in July and August.

12. Sam and Miriam Levering, "Seabed Mining," *Neptune* 14 (April 1979), 1, 4. In the same issue, Lee Kimball wrote an article entitled "Package" (p. 3) in an effort to help negotiators find potential trade-offs that might help complete the treaty.

Our most useful brainstorming in early 1979 was an all-day program on technology transfer in seabed mining in New York on Saturday, March 3. Organized by Lee Kimball, this program brought together several Third World delegates and UN officials to discuss issues relating to technology transfer with Prof. Howard Perlmutter of the Wharton School of Finance and other American experts. Perlmutter advised the delegates not to push compulsory transfer too hard, because success in this area depended on mutual cooperation and trust. A reluctant party could always manage to sabotage the transfer, especially because so much of the transfer would be not in machinery, but in the skilled personnel who knew how to make advanced technology work to accomplish specific tasks. The delegates and UN officials appreciated Perlmutter's commonsense approach and his avoidance of technical language. They also were pleased when Alan Kaufman, a lawyer for the high-technology firm Sedco, said that the company he represented would be open to working with the Enterprise if it were well organized and well run.[13]

Although we had several informative and well-attended seminars at Geneva in the spring, our most successful program at the conference in 1979 involved three simultaneous seminars that Lee organized on Saturday, July 21, just after the start of the resumed session in New York. One of these, which Tommy Koh asked us to arrange, focused on ways to provide more flexibility in the system of financial arrangements to meet the needs of developed as well as developing nations. In this seminar Koh, James Sebenius of Harvard, and Pali Kirthisingha of the UN Conference on Trade Development criticized the flat-rate system of taxing contractors that was part of the current negotiating text. Koh and others urged members of the G77 to accept U.S. requests for flexibility—a plea that bore fruit as the session proceeded. The second—led by a senior UN legal officer, Paul Szasz—dealt with decision making in the proposed International Seabed Authority. This seminar laid out options but made little progress toward consensus on this difficult subject.

13. Perlmutter's ideas are discussed in more detail in Burt Saunders, "Technology Transfer," *Neptune* 14 (April 1979), 3–4. In a letter I wrote shortly after the seminar, I commented that Kaufman "took down his hair and talked a lot of sense and had a way of charming the Third World delegates" (Miriam Levering to Dale and Cecile Andrew, March 8, 1979, Miriam Levering files).

The third seminar that day was on technology transfer. This program drew by far the largest audience, which pleased us because we believed that G77 fears that the Enterprise might never become operational were an obstacle to completing the negotiations. We felt that the delegates needed to know that technology would be available on the open market and that mandatory technology transfer probably would not be necessary.

One of the speakers, Ted Brockett of Sound Ocean Systems, Inc., commented that his company would be glad to license its technology for the Enterprise—or even put together a complete mining system. Other speakers, notably Louis Wells and Howard Perlmutter, stressed the need to promote positive interaction between suppliers and users of technology. The summary report on the meeting noted "a general feeling among participants that the Enterprise should be able to get into business, and that hardware for mining and processing operations would be available." The more difficult issue was "how to assure that the Enterprise would be equipped with the managerial skills in order to both (1) select technology and (2) put together and operate the systems" ("Summary Report . . . " 1979, 6).

We were pleased that roughly one hundred people, mostly delegates, attended these seminars. We were amazed when forty-five people, including many of those who had attended one of the seminars, came to a luncheon meeting the following Monday to hear three members of the Neptune Group summarize the ideas presented in the three programs on Saturday, and that twenty-six people attended another follow-up luncheon on Tuesday. The Neptune Group's efforts to educate delegates and UN staff on issues facing the conference were continuing to meet a real need.

Three weeks before these meetings, Koh, his wife, their two children, and his chauffeur spent five days visiting our orchard in southwestern Virginia. They stayed in a house close to ours where Sam had grown up, with a beautiful sixty-mile view of the front range of the Blue Ridge mountains and of valleys wending their way into North Carolina. Because the chauffeur was always immaculately dressed while Koh relaxed in blue jeans or jogging shorts, neighbors thought that he was the ambassador who was visiting the Leverings!

Koh ended up having several conversations about law of the sea during his visit, two of which were especially memorable. William C. Brewer of the Commerce Department, the main U.S. negotiator on financial arrangements, flew his small plane from Washington to the

Tommy Koh. Courtesy of the United Nations, photo 160, 768 by Y. Nagata.

modest airport in Mount Airy, North Carolina, the town nearest the orchard, to confer with Koh. Sam and I felt confident that Koh's and Brewer's discussion under a willow tree in our yard marked the first time that high-level negotiations had been conducted in our largely rural county.

At a dinner with friends in Mount Airy a couple days later, we asked Koh whether he sincerely opposed U.S. seabed mining legislation, or whether his public opposition was a negotiating ploy. He responded without hesitation that Elliot Richardson's support for U.S. legislation upset him greatly. He said that he could not understand why U.S. officials would want to make it more difficult for him to persuade his colleagues in the G77 to accommodate American interests. He had worked hard to obtain the navigation rights through straits and archipelagoes that the United States had wanted. If the mining legislation were passed, his influence in the G77 would decline. As the next chapter shows, I made good use of Koh's response in a conversation with Richardson a few months later.

During his visit Koh asked one favor: that Sam and I work to prevent the passage of seabed mining legislation until after the resumed session ended in August. Because we agreed with him that unilateral U.S. legislation should be postponed as long as possible, we found it easy to say yes. To be honest, we probably would have done almost anything Koh asked, for we were absolutely thrilled that the person whom we admired as much as any other living person would take time from his hectic schedule to visit us. For us, the Kohs' visit was a highlight not only of 1979, but also of our entire lives.

In retrospect, I wonder if my friendship with Koh and my desire to get a treaty may have made it harder for me to see that the details being negotiated might well be unacceptable politically in the United States and other Western nations. For example, I wrote approvingly in a letter to a Neptune Group colleague on September 12 that "the negotiators [in Koh's group] traded off easier financial terms for the companies with more developed nation payments to the Enterprise. This was the largest accomplishment of the resumed session" (Miriam Levering 1979). In other words, I was assuming that developed nations would agree to appropriate funds to subsidize the Enterprise, a competitor of private mining companies. At the time my enthusiasm for the negotiations clearly outweighed the concerns Sam and I had about whether a completed treaty could be ratified.

In 1980 the conference again met twice—in New York from March 3 through April 4, and in a resumed session in Geneva from July 28 through August 29. During these two sessions the conference continued to move toward consensus. We made significant contributions to changes that were made in two key areas: in the spring to changes that were made in Annex III, the section of the treaty setting up and defining the role of the Enterprise; and in the summer to agreement on decision-making procedures for the Council of the International Seabed Authority. Overall, our influence on the negotiations was at least as great in 1980 as in any other year.

The two stars of our conference-related work in 1980 were the experienced, widely respected Lee Kimball and a relative newcomer whom few delegates knew, Jim Magee. Recognizing that the structure of the Enterprise and its relationships with the ISA and with private and state mining companies needed to be more carefully delineated, Lee organized two small brainstorming meetings on the Enterprise during the month before the spring session, and then another, larger seminar for delegates and UN staff in New York on the Saturday after the session began. This last program was requested by Koh, who wanted our help in developing his next text on the Enterprise. Lee publicized the major conclusions of the two brainstorming meetings in a page-one article in the March 1980 *Neptune:*

> In one group the major concern was how the Enterprise could most efficiently and effectively learn to mine and conduct its operations. This group kept returning to joint venture arrangements as the best way for the Enterprise to acquire the necessary skills and technology to mine the deep seabed. . . . They opposed launching the Enterprise as a fully-staffed operation from day one and favored its organic growth toward a self-sufficient mining company. . . . The concerns of the second group centered around the importance of the Governing Board of the Enterprise being free from too much interference by the Council of the Assembly of the Authority in its day-to-day operations—whether decisions over finances or over modes of operation. (Kimball 1980b, 1)

Although Lee did her usual fine job of organizing brainstorming sessions and seminars at which new ideas could be brought effectively into the negotiations, her role as an informal, confidential bridge between negotiators representing different perspectives remained at least as important in early 1980 and beyond. Jim Magee's contributions were quite different: as the only member of the Neptune Group with a law degree and a willingness to tackle computer-based analysis, he undertook special projects for Koh and the UN secretariat in the late Carter years that helped move the negotiations forward. He was able to contribute in a major way in 1980 not only because he wanted to and because he had unusual analytical skills, but also because I succeeded in raising roughly $10,000 to support his work that year.

Jim was a unique member of the Neptune team. He was a sculptor of huge outdoor welded-steel sculptures and a designer for outdoor theater and opera. Having studied international law under famed Prof. Covey T. Oliver of the University of Pennsylvania Law School, he also supported stronger global institutions and took time off from his other projects to help us. To augment his income, he did welding jobs on oil rigs in Midland, Texas. It was dangerous work, with no protective railings. In the evenings he would draw on his legal training to analyze some portion of the negotiating text in great depth.

Jim also was unusual in that he was the only American member of the Neptune Group who came to us not through U.S. contacts, but by offering to volunteer with Quaker House in Geneva, which referred him to me. Eleanor Smith, a Neptune Group member in 1980 and 1981, recalled that Jim was "arrogant but very bright" (Smith 1989). Two other adjectives also fit: interesting and hardworking.

In 1979, at Koh's request, Jim had analyzed portions of the text on financial arrangements. Koh liked Jim's work, which Sam aptly characterized as "properly skeptical and extremely thorough" (Sam Levering 1980). In late 1979 or early 1980, Koh requested that we ask Jim to analyze the six pages of the negotiating text dealing with the Enterprise. Jim's memo in response ran seven single-spaced pages that included more than fifty questions designed to prod the negotiators to make the text less "disjointed" and "blurred" (Magee 1980a, 1, 7).

We sent a copy of Jim's memo to Koh. Together with our seminars and one-on-one work with delegates, it helped him effect some changes in the negotiating text on the Enterprise at the spring session. At a dinner at the Gold Coin Restaurant in New York on March 25, however, Koh regretfully told me that the Soviet delegates had not let him make the "radical changes" in the Enterprise that he had wanted to make.

The hottest issue remaining for the resumed session in Geneva involved the composition of, and voting procedures, in the Council of the ISA. Koh again sought Jim's help, and with modest funding from OEP, Jim moved into inexpensive lodging next to a noisy police station in midtown Manhattan that made it hard for him to get much sleep. He described his project in a letter to me on May 11:

> I will be working with Ali El Hussein and Mati Pal [of the U.N. secretariat].... I will be attempting to break article 161 down into its component parts, i.e., discovering which countries are eligible for Council representation as well as the attendant sub-issues, and then with the aid of a computer, analyze voting patterns on given sensitive issues. (Magee 1980b)

Jim worked hard to find current information about nickel and the other major minerals contained in seabed nodules—for example, rankings of all nations as producers and as consumers of them. A few weeks later, while cooking in my mountain kitchen, I got a call from Zuleta. "Could you find the money to put Jim Magee's figures on the computer?" he asked. "How much will it cost?" I replied. "About $3,000," he said. "Try not to go beyond that and I will help you," I answered.

I returned to the stove where the potatoes had burned. The phone rang again. It was a Quaker from New Mexico, Isabel Carroll, who had volunteered in our Washington office earlier. "The treaty must not fail," she said. "Of course it must not," I said. "I am sending you some money," she said. "Fine," I said, picturing $50 because I knew she was a retired teacher. "You'll get a check for $5,000 in the mail," she said. I was as stunned as I was grateful. There was my money for Zuleta plus $2,000.

A week later Mati Pal called. His voice was soft and gentle as usual: "Miriam, we need Jim Magee at Geneva to help us with the negotiations on the decision making in the Council." "Don't worry, Mati," I responded. "I have the money."

For the session at Geneva, Koh wanted Jim to use his and the computer's analytical skills to come up with fifty alternatives for decision making in the Council. Jim worked diligently for weeks, and finally laid out forty-five. "If Tommy wants five more, he will have to think of them himself," Jim told me. But the forty-five were ample: Koh and his fellow diplomats chose one of them and included it in the Draft Convention (informal text) that emerged from Geneva. The Neptune Group had made one of its most concrete and valued contributions.

When the resumed session ended in Geneva on August 29, there was widespread optimism among the delegates that, as a wire service story put it, "a final treaty will emerge from the next session next spring" (*Los Angeles Times* 1980, 12). Richardson fully shared this hope: "It is now all but certain that the text of a Convention on the Law of the Sea will be ready for signature in 1981" (ibid.). Because of the Neptune Group's emphasis on world law and governance, Sam and I were equally pleased by Richardson's prediction that historians would come to view the 1980 session of the conference "as the most significant single event in the history of peaceful cooperation and the development of the rule of law since the founding of the United Nations itself" (ibid.). Although we knew our work was not over, we spent much of the time smiling on our flight across the Atlantic.

Summing Up: The Group's Golden Years

Looking back, these were golden years for the Neptune Group. We worked closely with Richardson's team in Washington to help hammer out viable positions. Because of our extensive contacts with them, we knew we were working at the conference with a topflight U.S. negotiating team with enlightened yet realistic views of what the United States could reasonably accept. We appreciated and enhanced Richardson's ability to strengthen the conference moderates, which was essential if the negotiations were to succeed. During these years the U.S. delegation made friends with the moderates, strengthened their hands in the often heated discussions within the G77, and made many of them eager to obtain what their instructions required without totally selling their interests short. One of the moderates with whom we worked closely, Satya Nandan of Fiji, told me in March 1980 that the leaders of the G77 "would not have given the United States anything" if they had not liked and admired Richardson.[14]

Through our seminars and brainstorming sessions, we hoped not only to stimulate agreement on key issues such as financial arrangements, but also to make it possible for the G77 delegates and mining company representatives to get to know each other and to try to understand each other's worldview. Koh recalled that the seminars "helped us to understand problems, and helped us to explore solutions to problems. . . . There is nothing worse than negotiating in the absence of an agreed basis of facts" (Koh 1990). We also hoped that mining company representatives would come to appreciate the intelligence and ability of Third World moderates and the competence of the UN secretariat. In short, we hoped that the seminars would strengthen moderation in the conference and in the seabed mining industry by focusing on facts and the search for common ground.

Arthur Paterson recalled three main ways in which the Neptune Group became highly effective at the conference during the Carter years: the continued publication of *Neptune;* our success in "penetrating the emerging power structure of the conference," especially our close relationships with Third World moderates such as Koh and Nandan; and our service as an "honest information broker" by "helping to identify a problem and bringing in outside experts" (Paterson 1989). Underlying these contributions were the tremendous dedication and commitment of the members of the Neptune Group and the generous support given us by other NGOs, by foundations, and by United Methodist and Quaker bodies. Most remarkable, perhaps, were the hundreds of mostly middle-class Americans who wrote check

---

14. Nandan made this comment at a dinner at the Gold Coin Restaurant in New York on March 25, 1980, that he and Koh held in my honor.

after check to support an uncertain cause—a world governed increasingly by law—that was by no means assured but that they still believed in deeply.

During these years the negotiators moved closer and closer to agreement, and we were able to contribute more than we had ever thought possible to this process. It was a special, almost magical time.

6

Winning and Losing on Capitol Hill

WHEREAS THE NEPTUNE GROUP'S experience in working with the administration and at the conference during the Carter years generally was highly gratifying, our experience in the domestic politics of ocean law—centered on Capitol Hill and in the American news media—alternated between pleasing victories and disturbing defeats. Believing that successful negotiations required U.S. leadership, we sought above all during these years to defend the conference from strong nationalists in the U.S. business community, in the Congress, in the media, and among out-of-office conservative politicians who wanted the negotiations to fail unless the Third World delegates agreed to American conceptions of free enterprise in general and to virtually all of the companies' demands on seabed mining in particular.

Our greatest accomplishments on Capitol Hill during the Carter years came in helping our allies, notably Elliot Richardson and roughly twenty-five sympathetic members of Congress, make the seabed mining legislation increasingly internationalist and to delay its passage until June 1980, when it no longer caused significant damage to the negotiations. But we were not able to stem the steadily growing sentiment against the conference—even among allies like Sen. Claiborne Pell (Dem., R.I.)—that threatened to make a completed treaty even less popular on Capitol Hill than the narrowly ratified Panama Canal Treaty of 1977 or the unratified SALT II Treaty of 1979.[1] As Senate Foreign

---

1. The probable unpopularity of the treaty is captured in a letter that law-of-the-sea activist Robert Krueger wrote in 1978: "I think that we will get a treaty in the next 2–3 years and it will probably be as popular in the U.S. as the Panama Canal treaties" (Krueger 1978).

Relations Committee staffer Fred Tipson commented in February 1980, "satisfying the 'Group of 77' [at the conference] is pointless if we cannot also satisfy the 'Group of 67'—the number of Senators needed for advice and consent."[2]

Despite the dozens of speeches that Sam and I made to diverse audiences across the country, and especially despite the frequent, well-planned protreaty educational meetings involving Barbara Weaver and other Methodist Project staffers and hundreds of Methodist law-of-the-sea activists across the country, we were unable to expand our base of public support beyond the relatively small groups of church people, world-order internationalists, and environmentalists who backed OEP, UMLSP, the United World Federalists, and other protreaty organizations.[3] Although we and our allies won many battles on Capitol Hill, in other words, by 1980 we faced the grim prospect of losing the war: congressional and public support for completing and ratifying the treaty in the years ahead.

Our frustrations in the domestic political arena were like those faced by internationalist NGOs working on arms control and other issues in the late 1970s and early 1980s. The foreign-policy setbacks the United States experienced in the mid-1970s—the gasoline lines and sharply higher prices resulting from the new power of the Organization of Petroleum Exporting Nations (OPEC), America's defeat in Vietnam, Soviet/Cuban expansionism in the Third World, a new assertiveness among developing nations in international forums including UNCLOS III, and others—strengthened nationalist and conservative voices in America even before Jimmy Carter's election in 1976. When the Carter administration tried to pursue foreign policies based on internationalist conceptions of the U.S. national interest, therefore, it was subjected to harsher attacks from nationalist/conservative journalists and politicians than any previous post-World War II administration had experienced.

Beginning in early 1977 and continuing throughout the Carter years, conservative Republican columnists such as William Safire and Patrick J. Buchanan and conservative newspapers such as the *Wall Street Journal* offered sweeping criticisms of U.S. policy on law of the

---

2. These remarks from Tipson's speech on February 21, 1980, may be found in "Advice and Consent for a Law of the Sea Treaty," 2 (ML files).

3. For details about some of the Methodist Project's meetings around the country, see "The Nodule Onstage: Nationwide," *Soundings* 3:1 (March 1978), 4.

sea, and suggested at times that the U.S. withdraw from the conference.[4] After the fiasco involving Paul Engo at the conference in July 1977, even moderate-to-liberal papers such as the *Washington Post* joined in the attack. In an editorial on July 25, 1977, for example, the *Post* criticized "American apologists" for the Third World—presumably including members of the Neptune Group who frequently served as sources for *Post* news stories and wrote letters to the editor—and offered the "guess" that "few Americans are ready to 'pay' in foregone mining opportunities for a treaty that bestows in the future transit rights that, after all, the United States has enjoyed in the past without a treaty" (*Washington Post* 1977, 20).

In a balanced, analytical article in the *National Journal* in August 1977, Robert J. Samuelson noted the sharp split on the issue between internationalists and nationalists: "[T]he law of the sea has become—as much as anything else—a symbolic issue. Its supporters see it as a giant step toward world law and a gesture by the developed countries to share technology and wealth with poorer nations. Its critics see it as yet another instance when the United States allows itself to be tyrannized by hordes of developing countries into accepting things that are, basically, silly" (Samuelson 1977, 1343).

This growing polarization of views on law of the sea, combined with the increased defensiveness of many liberal internationalists in Congress, made our work there much more difficult during the Carter years than it had been under Presidents Nixon and Ford. We and our allies were fortunate—indeed lucky—to accomplish as much as we did.

Surprising Victories in 1977–1978

During the Ninety-fifth Congress (January 1977–October 1978), the nearly decade-long fight over seabed mining legislation reached

---

4. William Safire, "Strange Bedfellows," *New York Times*, January 27, 1977, and "Very Deep Thoughts," ibid., July 4, 1977; Patrick J. Buchanan, "Washington Parrots the U.N. Litany," *Greensboro (N.C.) Record*, May 30, 1977, 8; "Scuttle Law of the Sea," *Wall Street Journal*, July 22, 1977, 6. Safire's comment in his January 27, 1977, column was typical of conservative opinion: "The Carter-minded blocs [U.S. liberal internationalists in the State Department and elsewhere] want the seabed minerals for 'all mankind'; the U.S. should counter with a defense of free enterprise, free trade, freedom of movement on the high seas, and—not least—American national interests."

its highest intensity. Having seen the legislation die in Congress throughout the preceding five years, and needing a secure legal framework to maintain corporate funding for their projects, mining company officials were determined to get a mining bill passed during this session of Congress. "So many of the people involved depended on ocean mining for the future," Marne Dubs of Kennecott Copper recalled. "They desperately needed forward movement to keep their projects alive" (Dubs 1990). Conversely, Sam and I—and others who shared our perspective—were equally fervent in our belief that unilateral U.S. legislation at this time on the most difficult question facing the negotiators would be a major setback for the conference, quite possibly a fatal one.

Like most drawn-out political battles, this one featured colorful personalities as well as contrasting views of the issues involved. The nationalist side featured the outspoken representative of the U.S. Steel Corporation, Northcutt Ely, who denounced the negotiations at every opportunity. As Elliot Richardson was leaving Washington for his first negotiating session in 1977, for example, Ely reportedly said to him, "I hope you fail." The nationalist side also included two proudly patriotic, highly effective legislators—John Breaux (Dem., La.) and John Murphy (Dem., N.Y.)—who spurred the passage of the mining bill in the House in 1977 and 1978. According to Tom Kitsos, a legislative assistant to Breaux and Murphy on the House Merchant Marine and Fisheries Committee, Breaux "wanted to cut holes in some of Sam's arguments" and "get on the record that Sam's position was attackable" (Kitsos 1990).

The most colorful and controversial figure on the nationalist side was Leigh Ratiner, the former U.S. negotiator who, as an attorney in the Washington law firm of Dickstein, Shapiro, and Morin, worked as a lobbyist for Kennecott Copper from 1977 to 1979 and then fought the so-called Moon Treaty in 1979 and 1980. Ratiner was hardworking, determined, knowledgeable, shifty, and, as Elliot Richardson commented to me at the time, "effective." Dubs, who hired Ratiner to represent Kennecott's viewpoint, agreed that he was "an effective lobbyist and lawyer" (Dubs 1990).

Several supporters of the negotiations, in contrast, disliked Ratiner intensely. Lee Kimball, for example, remembered him as an "extreme opportunist" (Kimball 1989), and David Keeney of the Senate Foreign Relations Committee staff recalled that he was "a very abrasive personality, very aggressive" (Keeney 1989). Otho Eskin of the State

Department was unusually dispassionate: "He rubbed a lot of people the wrong way, but some liked his fire" (Eskin 1989).

Sam and I sided with the critics: although we agreed with Ratiner that the seabed mining text would have to be changed substantially before it could win Senate approval, we feared that his attacks on it, combined with his skillful lobbying on Capitol Hill on behalf of seabed mining legislation, would sink the conference. My contemporary correspondence suggests that, above all, Sam and I were concerned about Ratiner's clout. In April 1977 I wrote that he was "very persuasive and honey-tongued," and regretted that "the forces on our side" did not have the money to hire him (Miriam Levering 1977b). That September I commented that "Ratiner is very active and it will be difficult to keep him from getting what he wants" (Miriam Levering 1977c). And in February 1980, after hearing yet another of Ratiner's incisive critiques of the negotiations and the Moon Treaty, I summed up my view of him during the Carter years: "effective menace" (Miriam Levering 1980a).

The most colorful and outspoken figure on the internationalist side was Congressman Berkley Bedell (Dem., Iowa), a successful fishing-tackle manufacturer from Iowa and active Methodist who had won his seat in Congress in 1974, an election year in which popular anger about Watergate helped Democrats. Bedell's anger about the Vietnam War and the overall direction of U.S. foreign policy propelled him into politics, and also enabled him to connect with us for the first time when Sam was speaking against the war at a Methodist gathering in Des Moines in the early 1970s. Soon after the Bedells came to Washington in January 1975, Sam renewed his acquaintance and we became friends with Berkley and his wife, Eleinor.

Like many liberal Democrats and leaders in the mainline Protestant churches, Bedell believed that the Western nations needed to be willing to make sacrifices to strengthen world peace and improve living conditions in developing nations. More specifically, he believed that broad thinking about "getting the nations of the world working together more effectively"—not the narrow interests of U.S. mining companies—should determine U.S. policy on law of the sea. "I am more radical than Sam is," Bedell recalled. "I became very concerned about how the mining companies were calling the shots, and voiced it at all the meetings" (Bedell 1990). Bedell's brand of liberal internationalism did not please everyone. Depicting him as "not a serious figure," Dubs said that he had "very negative feelings about Berkley Bedell" (Dubs 1990).

Unlike Rep. Donald Fraser (Dem., Minn.) and other leading internationalists, Bedell was happy to play the often unpopular role of stubborn maverick rather than trying to gain long-term influence by being a conciliatory team player. "I wasn't influential," Bedell observed, "but I got a lot done because of my determination" (Bedell 1990).

To some on the nationalist side, Sam and I were as colorful—and as controversial—as Ratiner and Bedell. Not surprisingly, the main criticisms were that we were too idealistic and too sympathetic to developing nations.[5] Dubs recalled that a fellow mining representative, Richard Greenwald, was "psychotic" about the internationalist NGOs' influence (Dubs 1990). Another industry spokesman, Ely, once attacked me in a public meeting by saying that people should not listen to "little old ladies in tennis shoes"—a reference to the fact that I often wore canvas shoes with comfortable soles because of all the walking involved in attending meetings and visiting offices in Washington.

Although I had not planned to do so, I took revenge on Ely at a late afternoon meeting on law of the sea on Capitol Hill on November 2, 1978, sponsored by the University of Virginia's Center for Ocean Law and Policy. Early in the meeting, Ely seized the floor and, instead of asking a question as the format required, launched into a prolegislation, antitreaty speech and went on and on. I feared that he would consume the entire question period and shut out other questioners. As his oratory was in full flower, I felt the blood surging upward from my toes. My face became redder and redder. Suddenly I was surprised to hear my voice, usually so bland and conventional, rise loudly and clearly above his flowing prose: "Friend, thee has had thy share of our time."[6] Ely looked around, startled, his French cuffs slicing the air. A rustle stirred through the crowd. Ely added a sentence or two and then sat down.

The next day Sam and I heard that the incident had evoked amused comment at the State Department's law of the sea office, where Ely's implacable opposition to the treaty was well known. We did not hear

---

5. State Department official Myron Nordquist, for example, recalled hearing criticisms of Sam. Mining industry representatives "had difficulty understanding what his economic motive was" (Nordquist 1989).

6. This phrase is heard frequently in Quaker business meetings when a speaker is being verbose.

how Ely and other mining representatives reacted, but the incident may well have widened the breach between them and us even further.

Besides contrasting viewpoints and personalities, the battle over seabed mining legislation in 1977–78 involved complex, imaginative strategies and tactics on all sides. Partly because of Ratiner's and the industry's skill in pointing out the treaty's weaknesses and in warning of future "shortages" of critical minerals that could be alleviated only by seabed mining, the nationalist viewpoint was now the stronger one in Congress. But because most members of Congress still did not care very much about the issue, internationalists could slow down and "improve" the legislation in various ways. By insisting on the administration's right to control the bill's timing and most of its content, by using to his advantage the divisions in Congress between nationalists and internationalists, and by making ambiguous statements about the legislation that often gave comfort to both sides, Richardson proved to be the most capable strategist in the contest.[7]

Sam and his NGO allies in the religious, world-order, and environmental lobbies fought in 1977 and 1978 to prevent passage of any seabed mining bill that might hurt the conference. Sam worked especially hard in opposing a highly nationalistic bill, H.R. 3350, sponsored by Representatives Murphy and Breaux, and in helping draft and then supporting an internationalist substitute sponsored by Representative Fraser, H.R. 3652.

Sam and others from his office spent many hours meeting with key legislative assistants on these bills; he sent out several memos comparing the two bills; he testified frequently before congressional committees considering the two main bills; and he and I worked to get our supporters around the country to write to their representatives in Congress opposing the passage of any bill that might damage the conference. Kimball, Weaver, and other UMLSP representatives also worked hard to protect the negotiations. They were especially effective in getting Methodists in scores of congressional districts around the country to write or call their representatives in Congress to express opposition to nationalistic legislation.

In its various incarnations, the Murphy-Breaux bill had several features that the members of the Neptune Group strongly opposed.

---

7. Ratiner acknowledged Richardson's premier position: "The administration had control over the timing of it [the legislation]. Without Richardson's collaboration I couldn't have produced the legislation" (Ratiner 1989).

The worst of these, we believed, was a provision that would have granted U.S. seabed miners rights to mine specific sites on the ocean floor—a bold claim of control of land in the deep seabed that, by itself, might well wreck the conference. Among other things, we also opposed "investment guarantees" that would have required the government to pay the mining companies for any losses they incurred because of the treaty, the "grandfather clause" that would have required U.S. negotiators to achieve "substantially the same" terms and conditions for the miners in the treaty as were contained in the bill, and a "reciprocating states" provision that encouraged the U.S. government to negotiate a seabed mining agreement with selected nations outside the conference. Thanks to the work of internationalists such as Sam and Congressman Fraser and especially to Richardson's influence, these and other nationalistic provisions were deleted or weakened before the House passed a seabed mining bill in July 1978.

Whether one agreed with him or not, Sam was viewed as an expert on seabed mining and on the Conference. One of Sam's key assistants on the U.S. Committee, John McLean, recalled a meeting on the Senate side in the spring of 1977 that included, besides Sam and himself, a State Department official and several nationalist and internationalist legislative aides. The meeting was set up to discuss such key issues as "whether the legislation was going to move beyond the committee stage at this point" (McLean 1989). McLean remembered Sam's role: "Sam functioned in this meeting almost as a technical adviser. . . . Beyond being respected for the position he took, people knew that he . . . could give good, straightforward advice not only on his position" (McLean 1989).

The respect that members of Congress and their assistants had for Sam—and for Lee—also was reflected in the lengthy questioning that often occurred when Sam testified before congressional committees. When Sam appeared before the Subcommittee on Oceanography of the House Merchant Marine and Fisheries Committee on March 18, 1977, for example, Congressman Breaux entered into an extended discussion with him of his committee's bill, H.R. 3350, and of the likely reaction at the conference of its passage. Breaux argued that, although Third World delegates opposed any unilateral U.S. legislation, they did not "get bent out of shape" at the site-specific provisions of his bill. Sam responded that delegates who had seen the Fraser bill (H.R. 3652) preferred it to H.R. 3350 (*Hearings* . . . 1977).

From our standpoint, the battle over the mining bills during most of 1977 and 1978 bore considerable resemblance to a closely contested football game. I used this metaphor in describing the two key 1977 votes in the November issue of the U.S. Committee's newsletter:

> The Superlobbyists [Ratiner and Ely] had scored the first touchdown on August 9 in the House Merchant Marine and Fisheries Committee by passing nakedly "special interest" legislation, H.R. 3350. It gave U.S. companies unilateral permits to mine the international seabed before the . . . Treaty comes into force . . . [and] investment guarantee[s]. . . .
>
> Richardson's team scored the second touchdown. The vote on its key amendment in the House Interior Committee's marbled Room 1326 was close: 21 to 20. That October 26th amendment deletes the investment guarantees, which Richardson felt damaged the negotiations and the taxpayer. (*Sea Breezes* 1977)

In retrospect, an advantage of the football metaphor is that it fit the basic congressional division on seabed mining that solidified during late 1977 and early 1978: the nationalists, who enjoyed committed and effective leadership in both houses, and the internationalists, who largely followed Richardson's lead on what should be in the legislation and when it should be passed. A disadvantage is that it had no place for a third team, composed of a small number of fervent internationalists—notably Bedell and Rep. Ron Dellums (Dem., Calif.)—who supported the Neptune Group's view that no unilateral U.S. legislation should be passed in the near future because of the harm it might do to the conference. With the possible exception of Bedell, no one in Washington—certainly not Sam and I—thought that this obviously outnumbered and politically isolated third team could prevent the enactment of a mining bill if the administration wanted it passed in 1978.

Because we expected a bill to be passed, we worked with the internationalist team in the Congress and in the administration during the winter and spring of 1978 to make the legislation as consistent with the needs of the Conference as possible. In January and February, Sam focused on helping the staff of a subcommittee of the House International Relations Committee internationalize the Murphy-Breaux bill. We were very much pleased with the results, as I noted in a letter on 8 February:

> The House International Relations subcommittee marked up the mining bill by deleting the investment guarantee provisions, making revenue sharing mandatory before any mining permission is given, strengthening the statement in the bill that the U.S. is not assuming

sovereignty or sovereign rights to [the] seabed and a half dozen more improvements on Murphy-Breaux. (Miriam Levering 1978a)

In a letter published in the *Washington Post* on February 24, Sam noted that "Ambassador Elliot Richardson deserves much credit for aligning the Murphy-Breaux bill . . . more closely with the treaty"(Sam Levering 1978). That was certainly true, for it was Richardson's efforts far more than Sam's that led to the changes voted first by the subcommittee and then by the full House International Relations Committee in February. But Sam and other members of the Neptune Group also deserve some credit, for they worked closely with both officials and congressional staff to come up with specific changes in Murphy-Breaux. Another example of Sam's impact occurred in early March, when he phoned U.S. negotiator George Aldrich to alert him to the need to send a letter to Fraser to ensure that revenue sharing from mining income would be included in the bill that three House committees were putting together to take to the Rules Committee and then to the floor.[8]

Besides Sam, Lee Kimball of UMLSP was highly effective in 1978 in fighting Ratiner and the mining companies in Congress. Lee recalled one of her efforts to counter Ratiner:

> He put some amendments forward, and I read them and thought, "This would be a disaster." I did a paper real quick, and took it around to the offices of all the members of the House Foreign Affairs Committee, which I think stopped those amendments. That was one of those situations of knowing exactly what to do, when. The vote was going to be the next day on the floor, and you had to get something up there right then. Otherwise a lot of these people [members of Congress] would not quite realize the impact of what Ratiner was pushing, and would have accepted it. (Kimball 1989)

Lee made her largest and most memorable contribution to internationalizing the seabed mining legislation when the bill was being debated on the House floor in late July. Lee was deeply concerned about language in the bill that said that any treaty should "recognize the rights" of seabed miners to continue operating in "substantially the same way,"

---

8. The reference to Sam's phone call to Aldrich is in Miriam Levering to Barton Lewis and Bill Fischer, March 3, 1978. A copy of Aldrich's letter to Donald Fraser on revenue sharing, date March 23, was sent to the Leverings. Both letters are in the ML files.

and that the treaty should not "materially" impair investments. Because U.S. negotiators probably would not be able to persuade Third World delegates to agree to provisions that favored the mining companies so decisively, Lee feared that this language might lead the Senate to reject a more internationalist treaty if and when it was submitted for ratification. She thus prepared an amendment to soften the language.[9]

A major problem remained: with Fraser and other leading internationalists supporting the bill at this point, who would offer this amendment during the floor debate? While listening to the debate on July 25, Lee noticed that Rep. Millicent Fenwick (Rep., N.J.) made a comment "that led me to believe that she was sort of sympathetic to a different approach" (Kimball 1989). One of the interns in Lee's office that summer was a young man from a prominent family who had a young female friend who knew Fenwick. The rest of this remarkable story is best told in Lee's words:

> So I grabbed this guy [the intern] and said, "You've got to get us over there and get us into that office." And he called up whoever his friend was, or contact, and we actually got in to see Millicent Fenwick, who was smoking her cigar—her six-inch little Tiparillo.
>
> We came in, and she said, "Oh, you work for the Methodist Law of the Sea Project." I was with this guy, and I said, "Yeah." And she said, "Well, how come you aren't out there saving souls?" And I said, "Well the Methodist Church also works on social issues, policy issues."
>
> And then we got into talking about the substance of law of the sea, how it was important to get this treaty ratified, and so on. And I got out of there thinking, "Oh, Jesus, what a lightweight." She was all over the place, kind of flipping and flopping—that was useless.
>
> The next day I was not on the floor of the House when the bill finally passed. But I heard from somebody else that Millicent Fenwick had stood up, and in this kind of feminine, sad voice, said: "We shouldn't be too hard. Maybe we can do a few little changes here to soften this language. It couldn't do any harm."[10] And she got those amendments through that

---

9. These changes in the bill's language are described in Ann Pelham, "Seabed Mining Bill Awaiting Action by Senate Committee," *Congressional Quarterly,* August 5, 1978, 2073.

10. Fenwick's exact works were these: "Mr. Chairman, this is more in the form of a courtesy amendment, removing some of the buzz words that might offend other nations and make consultation and cooperation with them more difficult" (*Congressional Record* [House of Representatives], July 26, 1978, 7378).

totally undermined what the industry was trying to do. And I was so totally astonished that I almost fell off my chair. (Kimball 1989)

Despite Breaux's strong opposition, Fenwick's amendment—similar but not identical to Kimball's suggested changes—won internationalist support and passed. The Fenwick/Kimball changes were significant: the mild "allow" replaced "recognize the rights," the vague "similar" replaced "substantially the same way," and the ambiguous "unreasonably" replaced "materially" (Pelham 1978, 2073). The amended bill then passed easily, as expected, by a vote of 312 to 80 on July 26.

For us, the House's passage of the bill was both a victory and a defeat. We were pleased the legislation had been internationalized substantially and included the provision that the bill set up an interim framework for seabed mining that was to be superseded when the United States approved an international agreement. We emphasized the bill's "international" and "interim" characteristics in discussions with delegates and in press releases at the conference.[11] Passage was a defeat because it alarmed Third World delegates and appeared to hasten the day when the legislation would become law. During August and September we fervently hoped that Senate committees would internationalize the bill even further, and that the Senate would not pass the legislation before adjournment.

Just as July 26 was a good day for the Neptune Group in Washington because the House passed the Fenwick amendment, so August 17 was a bad day for several reasons. First, in testimony before the Senate Foreign Relations Committee, Richardson disagreed with two of our key positions: he said that he wanted legislation acceptable to the administration passed before Congress adjourned that fall, and he argued that passage would not hurt the negotiations. Second, two former U.S. negotiators on law of the sea, John Norton Moore and Richard G. Darman, testified before a subcommittee of the House Merchant Marine and Fisheries Committee that they disliked the current U.S. approach to the negotiations. Because the Neptune Group supported Richardson's work at the conference and wanted Congress to support him there, these criticisms were a blow to us as well as to him. And third, one of our longtime congressional friends on law of the sea, Sen. Claiborne Pell (Dem., R.I.), in effect told Sam during

11. See, for example, the Neptune Group's press release of August 17, 1978, which stressed the bill's "interim nature" (ML files).

his appearance before the Foreign Relations Committee that he needed to wake up and recognize the truth about the Third World's leadership at the conference. The exchange between Senator Pell and Sam documents Pell's anger at the time:

> Pell: What is the role that Paul Engo plays now at the discussions?
> Levering: That is one of the real problems, sir, and he is a great rhetorician.
> Pell: What is his responsibility?
> Levering: Unfortunately, he is still responsible for the work group, which is crucial, and there will be absolutely no progress as long as he is in charge.
> Pell: Exactly. He came here to Congress. We all knew nothing about it before and we had a very sympathetic outlook, and I think he turned off every man in the room. (*Hearings* 1978, 236)

The bad news kept coming. The Senate Foreign Relations Committee reported the seabed mining legislation in late August, and the Senate Finance Committee did likewise in early October. At the conference in New York in late August, a West German delegate informed me about a conversation in which Richardson had told him that he expected President Carter to sign the legislation before November 20 (Miriam Levering 1978b). The reaction to the legislation among many Third World delegates was extremely negative: some Third World radicals accused moderates of allowing the bill to pass and called for a resolution in the General Assembly condemning the United States. And the letters that we and our supporters around the country wrote to members of the Senate in September, urging them not to pass legislation that fall, generally were met with polite but firm disagreement with our position. Sen. Charles Percy (Rep., Ill.) expressed a typical internationalist opinion: "Though I am not unsympathetic to your arguments about the effects of this bill on the Law of the Sea Conference, I have decided to support its passage as well as to continue to support an internationally negotiated solution in the long run."[12]

---

12. Charles H. Percy to Barbara Weaver and Samuel R. Levering, September 25, 1978. For a similar response by another internationalist Republican, see Jacob Javits to Richard H. Post, October 3, 1978. For a response from a nationalist Republican that emphasized shortages of minerals as the main reason to pass the bill, see John Heinz III to Palmer H. Futcher, October 3, 1978. These and other letter from senators—also disagreeing with the Neptune Group's position—are in ML files.

With the administration and almost all Senators supporting the legislation, Sam and I assumed that it would pass that October. For once in our lives, we were thrilled to be wrong. The persistence of Berkley Bedell, plus the ability of one senator to block a vote on a bill at the end of a session, quashed the mining bill for at least a few more months.

The credit for preventing the passage of the legislation belongs primarily to two liberal mavericks: Bedell, a representative from northern Iowa, and James Abourezk, a Democratic senator from neighboring South Dakota whose service in Congress ended that fall.[13] Bedell made several phone calls to Abourezk about the legislation; "he was quite adamant each time he called me about holding fast," Abourezk recalled (Abourezk 1983). Bedell also flew to South Dakota to visit Abourezk to ensure that the senator would not waver under pressure (Bedell 1990). With Bedell's encouragement, Abourezk decided that he would threaten to filibuster if necessary to prevent the bill's passage.

The Bedell/Abourezk strategy worked perfectly. Abourezk put a "hold" on the bill and told Senate Majority Leader Robert Byrd (Dem., W. Va.), who was working to complete the Senate's business, that he was determined to prevent it from passing. Calling from Europe, Richardson pleaded with Abourezk to release the bill, but he refused. Richardson remembered arguing with both Abourezk and Bedell about their stand (Richardson 1990). Besides the pressure from Richardson, Abourezk recalled Congressman Breaux "walking over to the Senate floor and asking if I would let the bill go through" (Abourezk 1983).

To our amazement and relief, the "third team" of fervent internationalists had scored the final touchdown on law of the sea in 1978. But any celebrating we did was muted by the awareness that the bill could pass quickly in the next session of Congress, and that the ultimate success or failure of the treaty in the U.S. political arena almost certainly depended on the fate of Richardson's brand of internationalism, not on Bedell's and Abourezk's.

Uphill Struggles in 1979–1980

The years 1979 and 1980 brought us yet another set of often surprising wins and troubling losses. On the positive side, we helped delay the passage of the legislation for much longer than we had

---

13. Schmidt 1989, 94–95, also analyzes this incident. Schmidt notes that Abourezk was "one of the Senate's most liberal members."

expected, and to make it even more internationalist and protective of the ocean environment than it had been when the Senate failed to pass it in October 1978. More negatively, support for the negotiations in Congress and among American business leaders continued to decline and the concept of the "common heritage of mankind" came under blistering attack. By 1980 it appeared far more likely that the negotiators would complete the treaty than that the Senate would ratify it.

Sam and I fully expected the new Ninety-sixth Congress to pass a mining bill, but we still thought that we might have some influence on its timing and especially on its content. "For a long time there was a question of whether there'd be a bill at all . . . ," Sam told a *Christian Science Monitor* reporter in March 1979, "but now the question is what kind of bill. . . . We'd like to see the conference have at least three more years to work before even exploration begins" (*Christian Science Monitor* 1979, 9).

Sam suggested in early 1979 that the new House bill, H.R. 2759, be amended to allow no exploration before January 1, 1981, and no commercial recovery before January 1, 1985. To give additional substance to the principle that all nations owned the minerals, Sam also proposed an amendment that would require revenue sharing to be raised from 3.75 percent to 10 percent of the imputed value of the minerals mined. In the bill that was finally passed in June 1980, Sam and other internationalists won on the first point—indeed, the date of commercial recovery was pushed back even further, at Richardson's insistence, to January 1, 1988. But they lost on the second: the 3.75 percent for revenue sharing remained unchanged.[14] Sam also argued for—and helped secure—stronger environmental provisions in the legislation. Finally, Sam repeatedly urged Congress to postpone passage of the bill until after the next session of the Conference to avoid harming the negotiations. Because many members of Congress were fed up with the slow pace of the negotiations, and because Sam had used precisely this argument for several years, it probably had little effect in 1979 or 1980.

Our friendships with Koh and Richardson may well have paid off for us by delaying the passage of the legislation for six or seven months—from November or December 1979 until June 1980. Dur-

---

14. For a summary of the provisions of the 1989 law, see Schmidt 1989, 97–98.

ing a recess in hearings before the House Foreign Affairs Committee (its name had changed back again) on November 1, 1979, Richardson said to me, "Does Tommy Koh really oppose [the passage of] this legislation, or is he using this [his public opposition] as a bargaining chip [negotiating tactic]?" I responded, "He's not using this as a bargaining chip, because otherwise he never would have talked about it in Mount Airy, North Carolina, with our friends."

I explained to Richardson that Koh, during his visit to our orchard the previous July, had said at a dinner party in Mount Airy that the United States, by passing the mining bill, would make it harder for him to persuade his Third World colleagues to accommodate the United States on seabed mining. He also had said that he had done much to help the United States on guaranteed passage through straits and other issues, and that he therefore could not understand why Richardson would want to undercut him in this way. Richardson nodded sympathetically as I told this story. When the recess ended, Richardson disappeared into a back room with the committee and, I learned later, told them not to pass the mining bill that year.[15]

At this point Richardson apparently decided that he wanted neither the House nor the Senate to pass the bill in 1979. About this time he told Abourezk during a chance encounter in an airport that, in Abourezk's words, "he had changed his views on the legislation and had come around to agree with my point of view that it would be destructive of the treaty" (Abourezk 1983). But Richardson did not have total control of the situation. In early December Sam learned that President Carter recently had

> decided not wholly in Richardson's favor in the conflict between State, Defense and National Security Council who were on Richardson's side and the Treasury, Interior, Commerce, and Office of Management and Budget on the other. He decided to let the bill pass one house, namely the Senate, and not the House until after the March session in New York. Richardson is afraid the momentum may carry it through the House where bills can slip through more easily. (Sam Levering 1979)

With approval from the administration, therefore, S.493 passed the Senate easily on December 14, 1979.

15. During an interview a decade later, Richardson nodded as I retold the story and said that he had indeed "held it up in the House for one year" (Richardson 1990).

Because we knew that the legislation was being held up in the House, Sam and I viewed Senate passage as only a small setback. Of much greater concern was the unprecedented criticism of the conference and of the U.S. negotiating position that emerged in the debate, even on what we had thought was the relatively friendly internationalist side. In a careful study of the statements fifteen senators presented on the floor, Ken Short, one of our interns, found that six "were distinctly opposed to the treaty in any form it is likely to take soon" and the other nine, though "more supportive of the treaty," warned that a treaty that could be ratified had to give U.S. mining companies basically what they were asking for: protection for their investments, assured access to mining, security of tenure, no forced transfer of technology, and no unreasonable financial burdens. Moreover, the Senate would not ratify the treaty unless the voting system in the Council of the International Seabed Resources Authority protected U.S. interests. Only one senator, Paul Tsongas (Dem., Mass.), opposed the bill along the lines we had been stressing; namely, that it might well harm the negotiations (Short 1980, 4).

Two months later, after talking with staffer Fred Tipson of the Senate Foreign Relations Committee, Sam wrote a rare letter to Richardson to "bring to your attention the urgency of high level law of the sea work on the Hill." At the present time, Sam noted, paraphrasing Tipson, "support of a majority of the Committee is not assured, and there is no certainty that even one really influential senator will put forth a major effort for ratification." Sam had been upset since 1977 by Richardson's failure to appoint an experienced administration lobbyist to counter the mining industry's influence; he now suggested that William Brewer of the Commerce Department be assigned to "discuss U.S. ocean options, including the Treaty, with members of Congress and staffers."[16] Although Sam's proposal had merit, it probably was too late to bring about a significant change in congressional attitudes toward the negotiations before the 1980 election.

---

16. Sam Levering to Elliot Richardson, February 26, 1980 (ML files). Lee Kimball also was aware of the administration's poor work in lobbying Congress—and in keeping sympathetic members of Congress informed—on law of the sea. Kimball recalled that Fred Tipson "once called our office in New York and noted that the administration's Congressional relations staff never responded to him and that their information was never as good as what he could get from us" (Kimball 1995).

Besides the declining support for the negotiations in the Senate, another development in the fall and winter of 1979–80 upset and discouraged us. This was the political fate in the United States of the "Agreement Governing the Activities of States on the Moon and Other Celestial Bodies," commonly known as the Moon Treaty. The UN Committee on the Peaceful Uses of Outer Space completed the little-publicized negotiations on this treaty on July 3, 1979, and it was approved by the UN General Assembly and opened for ratification in December. U.S. diplomats had taken the lead in moving negotiations on this treaty forward in the early 1970s, and had approved the use of the phrase "common heritage of mankind" to describe the resources on the moon and other celestial bodies. Because of the difficulties in the law of the sea negotiations, that phrase had lost its luster by the late 1970s. Moreover, in the summer of 1979 a little-known group containing approximately thirty-six hundred American space enthusiasts, the L-5 Society, hired Leigh Ratiner (recently dropped as a seabed-mining lobbyist by Kennecott Copper) to spearhead a congressional and media campaign against the Moon Treaty.[17] Ratiner later told me that the L-5 Society paid him $100,000 to defeat the treaty and that he had taken the job because he "needed the money." Much to our dismay, the organization got its money's worth.

Ratiner's persuasiveness and his access to Washington's corridors of power paid off handsomely on October 30, 1979. That day a long article on the Moon Treaty and the L-5 Society's opposition to it—including several quotes from Ratiner—appeared in the *Washington Post* (Dewar 1979, 3). Of much greater significance, the chairman of the Senate Foreign Relations Committee, Frank Church (Dem., Idaho), and the ranking minority member, Jacob Javits (Rep., N.Y.), sent a lengthy, Ratiner-inspired letter to Secretary of State Cyrus

---

17. According to a contemporary article, the L-5 Society was "formed at a 1975 Princeton conference on space manufacturing facilities and named for an Earth-orbit location that is considered well-suited for human colonization" (Dewar 1979, 3). The group's opposition to international (or even national?) regulation of activities in outer space was summed up in a comment made by the outgoing president, Carolyn Henson of Tuscon, Arizona: "For those of us who plan to go into space, it's a give-me-liberty-or-give-me-death kind of issue" (quoted in Dewar 1979). The figure of thirty-six-hundred members is contained in Nossiter 1980, 8E.

Vance that urged the administration to withdraw its support for the Moon Treaty. Church and Javits argued that the "set of draft treaty articles now before the [law of the sea] Conference sets forth an interpretation of the 'common heritage' which does not conform to the national interests of the United States or of other countries with free enterprise/free market economies. . . . " Linking the two negotiations, the senators said that they were "skeptical of further efforts to extend the concept of the common heritage when the understanding of this principle on the part of many countries of the world is so contrary to our own interests" (Church and Javits 1979).

With Sam's and my encouragement, Lee Kimball and her colleagues at the Methodist project took the lead in trying to counter Ratiner's new thrust. Seeking to refute the senators' arguments, she sent a letter to Vance on November 8. She contended that the senators' "allegations belie the necessary cooperation which any development of resources beyond national jurisdiction entails. . . . An element of cooperation and regulation is necessary to protect the interests of all those who would exploit them" (Kimball 1979b). Lee and her Methodist colleagues continued to work on this issue through June 1980, long after Ratiner and the two senators had prompted the administration not to sign the Moon Treaty and to set up an interagency committee to review U.S. policy in this area.

That Kimball and her colleagues were outmatched financially—and probably politically[18]—became clear when one of the nation's giant defense contractors, United Technologies, used paid advertisements to attack the Moon Treaty in the *Washington Post* and other newspapers. In the typical ad that appeared in the *Post* on February 14, 1980, United Technologies linked the Moon Treaty and the law of the sea negotiations, and argued that the "concept called 'the common heritage of mankind' . . . is intended to place seabed mining technology and resources under the control of countries seeking to bring western nations to heel." These "so-called non-aligned nations, guided by the Eastern Bloc" were seeking "redistribution of the world's wealth" (United Technologies 1980, 2A). Such intensely nationalist

---

18. Margaret Gayley, a staffer for the House Foreign Affairs Committee, recalled that "Sam and Miriam and Lee were not as effective in supporting the Moon Treaty as Leigh Ratiner was in criticizing it. There was a different set of players on the moon, and it takes a while to rev up your contacts. The industry people had a ready set of contacts" (Gayley 1990).

and conservative attacks became commonplace in criticisms of the law-of-the-sea negotiations after Ronald Reagan became president in January 1981.

The swift and decisive defeat of the internationalist viewpoint on the Moon Treaty, combined with the sharp Senate criticisms of the law of the sea negotiations, made it clear to Richardson and to others supportive of the conference, including the members of the Neptune Group, that the protreaty side would need to undertake a major effort to build support for ratification. Barbara Weaver of UMLSP took the initiative while the Moon Treaty episode was still in its early stages. She set up a meeting with Richardson in September 1979 to discuss laying the groundwork for coalition building for the treaty. Richardson agreed with Barbara that it was time to get started, and told her that it needed to be large-scale and include well-known Americans from both political parties. Richardson said that he had planned to get started in early 1977, but the Engo fiasco and the fight over the legislation had interfered.[19]

Planning for a new, protreaty organization began in earnest in late 1979; it took shape during several organizational meetings in early 1980 as Citizens for Ocean Law. Formed largely on the basis of Richardson's thinking and under his direction, COL featured Washington insiders/friends of Elliot—Melvin Conant, Cecil Olmstead, and Charles Maechling Jr.—as chairman of the board, treasurer, and secretary, respectively. Other prominent people on the board included Paul Fye of the Woods Hole Oceanographic Institution, John Temple Swing of the Council on Foreign Relations, and Russell Train of the World Wildlife Fund.

The Neptune Group, which Richardson admired, also was well represented: Barbara Weaver, A. Barton Lewis, and I served on the board and on important committees, and Lee Kimball soon was working half-time as the executive director. We fully supported COL's "emphasis ... on the interdependence of nations and the importance of international accords to further those interests," its desire to "establish working relationships with a wide range of existing associations," and its intention to "arrange briefings, programs and publications for congressional offices and media outlets" ("Citizens ... " n.d., 1, 4–5).

19. Weaver's meeting with Richardson is described in a letter from Miriam Levering to Barton Lewis and William Fischer, September 28, 1979 (ML files).

All of us hoped that COL would be able to attract the big-name supporters and the large-gift donors that were not likely to become involved with the Neptune Group organizations. If COL could become even half as respected on law of the sea as the Committee on the Present Danger and Americans for SALT were on arms control, perhaps our new group could tip the balance in favor of ratification.

Sam and I were well aware by 1980 that ratification would be an uphill fight, and that the most effective opposition was likely to come from the mining companies and their allies in American business. In an effort to increase communication between mining representatives and protreaty internationalists, Sam and I met on April 21, 1980 with Robert Ferguson, the executive vice president of U.S. Steel.

Unlike some in his industry, Ferguson told us, he did not wish Richardson "bad luck" in his effort to negotiate a treaty. But he said that he was "adamant" in his refusal to accept the treaty's technology transfer provisions—a point that I mentioned in a subsequent letter to Richardson in the hope that he could get these provisions eliminated or at least softened substantially (Miriam Levering 1980b).

By the spring of 1980 Sam was eager to devote more of his energy to the orchard. He also felt that, given Richardson's determination to get the legislation passed by early summer and strong support for this position from both nationalists and internationalists in Congress, there was nothing we could do to delay passage further. Sam also was pleased that the Methodists had developed over the years an extensive lobbying program on Capitol Hill on behalf of the treaty. Accordingly, Sam ended the work of the U.S. Committee about the first of June. Although he continued to serve on the Public Advisory Committee, he was less active on ocean issues than I was over the next few years.

A Balance Sheet for 1977–1980

Despite the growing difficulties on Capitol Hill from 1977 forward, Sam and I were delighted that the seabed mining legislation was postponed until June 1980, and that the final legislation was so internationalist that it did not harm the negotiations. We believed that the Neptune Group deserved some credit for both of these developments. But we were disappointed that, despite our work and the efforts of other internationalists, support for the negotiations in Congress and in the U.S. business community declined steadily during these years. We also were disappointed when high officials in the Carter administration decided not to seek Senate approval of the Moon Treaty in late 1979 and early 1980 and made only a muted defense of the "common heritage" principle that it embodied.

Overall, our experiences in Washington during the Carter years did not discourage us, because the administration generally supported the negotiations and because we won at least as many battles on Capitol Hill as we lost. Only after the Reagan administration made fundamental changes in U.S. policy did we truly know what it meant to feel discouraged.

7

Seeking to Counteract the Reagan Shocks

LIKE THE SUDDEN BUT NOT UNEXPECTED earthquakes that frequently jolt the conservative president's home state of California, the first two Reagan shocks to UNCLOS III occurred in early March of 1981. On Tuesday, March 2, with Secretary of State Alexander Haig out of town, Deputy Secretary of State William Clark convened the Senior Interagency Group on Law of the Sea. The purpose of the meeting was to decide whether to continue with serious negotiations (and with experienced U.S. negotiators) at the upcoming session in New York, or to suspend active U.S. participation in the negotiations while the new administration conducted a thorough policy review. Although acting head negotiator George Aldrich argued that the United States should continue to seek an improved treaty, most of the participants including Clark favored the second option. An important statement was released to the press at the end of the meeting:

> [T]he Secretary of State has instructed our representatives to seek to ensure that negotiations do not end at the present session of the Conference, pending a policy review by the U.S. Government. The interested departments and agencies have begun studies of the serious problems raised by the Draft Convention, and these will be the subject of a thorough review which will determine our position toward the negotiations. (Schmidt 1989, 223)

The second shock came just five days later, on Saturday, March 7. On that day Clark told Aldrich that he and most other officials working on law of the sea were being reassigned because the administration wanted a "clean break with the past" (Schmidt 1989, 225). A much smaller U.S. team replaced them, led by as yet unconfirmed assistant secretary of state James Malone, a newcomer to the intricacies of the negotiations. Over the next couple years Malone struck Sam and me and others who shared our perspective as ineffective and uninformed,[1] not only compared with his predecessor, Richardson, but also judged against the hard-nosed bureaucratic infighter who handled the day-to-day details of the policy review, Deputy Assistant Secretary of State Theodore Kronmiller, a longtime aide to Sen. John Breaux and persistent critic of U.S. policy in the negotiations. Like several other high officials, including Secretary of the Navy John Lehman, Kronmiller feared that the Soviets and their allies in the Third World might try to cut off U.S. access to strategic minerals—including, through UNCLOS III, access to deep seabed minerals.

We also thought that Malone compared poorly with the person the administration with some misgivings chose to be the de facto head negotiator, the knowledgeable, mercurial Leigh Ratiner. Always far more internationalist than ideological nationalists such as Kronmiller, Ratiner switched back to being protreaty on the premise that, as a key negotiator once again, he would be able to improve the treaty sufficiently to make it serve the U.S. national interest. But perhaps we were unfair to Malone: admiring (almost worshiping) Richardson and Aldrich and disagreeing strongly with Ronald Reagan's policy, we almost certainly would have been critical of anyone whom the openly conservative, nationalistic president and his top advisers would have appointed to head the U.S. delegation.

The members of the Neptune Group reacted to these two shocks—and to the other major one that occurred in April 1982—with anger, with our usual intense activity, and with disappointment and frustration. After recovering from the initial shocks, we also exhibited considerable creativity in an ongoing effort over the next two years to save the treaty.

My anger came through clearly in a letter to supporters on March 4, shortly after I had learned of the review: "We are convinced that we must express our indignation and outrage to the new administration, which should have continued the negotiations in good faith" (Miriam Levering 1981a). Anger and bitterness that the negotiations had been stalled also appeared in a letter I wrote two weeks later from the conference in New York: "The U.S. delegation got lambasted yesterday by

1. "Malone, in my view, is worthless," Richardson recalled. "What really turned me off on him was, he came in having no knowledge [of law of the sea]" (Richardson 1990).

most other nations at the General Plenary. Can't blame them" (Levering 1981b). Lee Kimball's anger and her close ties with many delegates came through in an editorial she wrote that spring for *Soundings:* "Does the United States favor a comprehensive international oceans agreement, or not? Does it continue to support certain fundamental principles in that agreement, or not? UNCLOS delegates from many nations want some assurances from the U.S." (Kimball 1981a, 5).

Besides the usual seminars and receptions at the conference in New York (March 9–April 24) and stepped-up contacts with sympathetic U.S. officials and congressional staffers in Washington, the intense activity involved far more contacts with journalists than ever before. Lee spent the first few days after the announcement of the review answering reporters' questions and alerting them to the importance of the story. Partly because officials in the new administration were less willing to talk openly with reporters about U.S. policy, and partly because the conflict and confrontation among Americans knowledgeable about law of the sea and between the administration and the conference increased the story's newsworthiness, journalists contacted Lee and other Neptune Group members repeatedly in the ensuing weeks. As I wrote to a friend on April 6: "We have been swamped with work in New York because the media has given the law of the sea more attention in the last month than it had in the previous seven years! Consequently our New York phones have been constantly ringing" (Miriam Levering 1981c).

Above all, the shocks that occurred in March 1981 and thereafter disappointed and frustrated us. We were disappointed and frustrated that, just when the treaty appeared to be nearly finished and our able friend Tommy Koh was elected president of the conference, the United States and other Western industrial nations might refuse to help conclude the negotiations and sign the treaty. We were disappointed and frustrated that, unlike previous administrations, the new one was listening to the mining representatives' and conservatives' criticisms of the draft treaty much more than it was listening to people like Richardson and us.[2] Indeed, we often doubted that it was listening

---

2. The chief U.S. negotiator, James Malone, recalled that he would have appreciated less involvement by Elliot Richardson on law of the sea in 1981 and 1982. He thought that Richardson did not really accept the conservative shift reflected in Reagan's victory in the 1980 election. Malone also said that he was irritated by some of the positions taken in publications by the Neptune Group and by environmentalists (Malone 1994).

to us at all. And we were disappointed and frustrated that, despite years of effort by multilateral internationalists like ourselves to build support for the treaty, it seemed highly unlikely that there would be sufficient public and congressional pressure to force Reagan to change his basic approach, often called unilateral internationalism.

Although staff members and volunteers from the U.S. Committee (which Sam briefly revived) and especially from the Methodist Project continued to lobby on Capitol Hill on behalf of the treaty, Sam and I were well aware that Congress could make a significant difference on this issue only if Republicans (now a majority in the Senate) and conservative Democrats pressured the administration on behalf of the Pentagon's interest in the treaty's provisions relating to navigation. Jim Magee of the U.S. Committee and Lee Kimball and other Methodist representatives worked hard on the Hill in May–June 1981, stressing U.S. gains in navigation under the treaty and apparently helping persuade conservative Sen. Henry Jackson (Dem., Wash.) to write Malone on this issue.[3] Unfortunately, Reagan's Defense Department was itself divided on the merits of the treaty, and thus was much less supportive than it had been earlier. Overall, we were disappointed—though not surprised—that relatively few members of Congress openly challenged the popular new president's policies on law of the sea.[4]

Although disappointed and frustrated, we were determined in 1981 and early 1982 to help the negotiators achieve the best possible treaty, one that the United States might sign and ratify either then or

---

3. Henry Jackson to James L. Malone, June 23, 1981, (ML files). Jackson stated his concern that the developing countries might agree to a treaty without the United States, and added: "I am also not entirely satisfied that the national security interests relating to navigation and overflight which were the original impetus for U.S. involvement in the law of the sea process are being given the weight I assume we still think they deserve vis-a-vis deep seabed mining." Aware of Jackson's clout and needing support from any quarter, we were pleased by his letter and by some others that we stirred up.

4. One reason for the relative quiescence of Congress was many Democrats' reluctance to challenge Reagan because of the new president's popularity. "On this issue, on the U.N. in general, and UNESCO in particular, the Democratic Party let the nation down," liberal representative Jim Leach (Rep., Iowa) recalled. "When the Reagan administration went off the deep end on foreign policy, the Democrats should have offered an effective opposition. But, except for Bedell, the Democratic Party did not take notice on U.N. issues.... There was fear of a backlash" (Leach 1990).

after the nationalistic fires of the early 1980s had died down. Even after Reagan announced on July 9, 1982, that the United States would not sign the treaty, we hoped that the administration would participate in a continuing effort to improve it—or that at minimum it would not support legislation that contradicted many of the treaty's provisions. In short, there was too much important unfinished business to permit disappointment and frustration to curtail our efforts before we felt confident that, at least for the early Reagan years, we had done all that we could do.

Working to Save the Negotiations

Except for a few months after the Engo shock of July 1977, we had assumed from 1973 forward that the U.S. government's main goal in the negotiations was to complete a treaty, one that necessarily would contain compromises and trade-offs with Third World and other interests. Thus the negotiations were paramount, and we put our greatest efforts into facilitating agreement at the conference. In 1981 and 1982, however, the Reagan administration's policy review took center stage. Not only did progress in the negotiations come to a virtual halt at the sessions in the spring and summer of 1981, but it was not even clear until the following winter that the United States would return for the final session in March-April 1982. Indeed, the review did not end officially until President Reagan issued a statement on January 29, 1982, indicating that the United States would return to the final negotiating session and would seek the achievement of six specific objectives.[5] Even after the president's statement, conflict continued within the administration between those such as Haig and Ratiner who believed that an acceptable compromise treaty could be achieved through hard bargaining, and those such as Clark and White House adviser Edwin Meese who thought that only G77 acceptance of virtually all U.S. demands could lead to a treaty that the United States should consider signing.

Given the centrality of the policy review and given the Neptune Group's belief that a treaty without U.S. participation would be much weakened, our work between March 1981 and January 1982 necessarily focused on the policy review. Specifically, we urged that the

---

5. For details of the review and its effect on the negotiations, see Schmidt 1989, 228–42.

Miriam Levering working at the OEP office in Washington in early 1980s. Courtesy of Miriam Levering.

United States should return to serious negotiations as soon as possible—hopefully at the August 1981 session—and that it limit its demands to a short list that the conference might well be able to accommodate. We feared that too many demands would cause the delicate compromises embedded in the treaty to unravel. We also encouraged the agencies making the review not to focus entirely on seabed mining, but to consider the consequences of a failed treaty to other U.S. interests, including navigation and relations with the Third World.

The Neptune Group members who made the greatest efforts to influence U.S. policy in 1981 were Lee and Sam. Both made important contributions at the Public Advisory Committee meetings on June 8–9, and both offered suggestions thereafter to key U.S. officials. At the advisory meeting Lee urged Malone and Kronmiller to "develop some kind of outline, at least, of a restructured Committee I package" to present to other delegations at Geneva in August. A specific proposal would have two advantages: it would "give other delegates more to get their teeth into," and it would "indicate to other countries that there is a will on the part of the United States to negotiate a treaty." Malone responded that the U.S. delegation could "explore . . . the various alternatives that would make up the package without actually attempting to negotiate these things." He felt confident that "the reaction that we get" would make it possible to know "where we can then go from there and the type of advice that we can present to the President" (Public Advisory Committee 1981a, 104–6).

Sam and several other members of the advisory committee agreed with Lee that the United States should negotiate seriously at the August session. Sam perceived a "crystallization of intransigence . . . in a considerable part of the Third World," and argued that it would be "better if we take this opportunity to move ahead rather than alienate people by stalling." Sam urged Malone and Kronmiller to develop a "pretty complete" list of "changes needed," and then specified some that he thought were necessary: more definite guarantees of access to mining sites, deletion of provisions requiring mandatory transfers of technology, and an "assured seat" for the United States on the Council of the ISA. Sam concluded by arguing that these changes could be achieved, but only if other delegations "recognized that we are really trying to get an acceptable treaty" (Public Advisory Committee 1981a, 122–24).

Lee followed up with a letter to Malone (with a copy to Kronmiller) in which she repeated her "endorsement for taking some proposals to the Geneva session . . . on the basis of which the U.S. delegation could initiate negotiations." As an example of a possibly fruitful proposal, she suggested that the United States might be able to "eliminate the technology sale obligations and reduce national obligations to finance Enterprise operations" if joint ventures between the Enterprise and mining companies were "re-explored" (Kimball 1981b). In a memo to Kronmiller, Sam fleshed out this idea, arguing that the treaty should specify "that at least the first two mining operations by the Enterprise shall be *joint ventures,* voluntarily agreed upon with companies, unless the Enterprise can obtain *voluntary financing.*"[6]

Sam met with Kronmiller in July to discuss his and Kimball's proposals, and Kronmiller thought enough of these ideas to send a copy of Sam's memo to Malone in Geneva even though Malone had had

6. Sam Levering to Ted Kronmiller, July 1, 1981 (ML files). Emphases in original.

a chance to read it before leaving Washington.[7] In a letter in early August, Sam was cautiously optimistic about recent trends in U.S. policy: "They [Malone and Kronmiller] have been welcoming ideas and input from us and our present relations with Ted Kronmiller are good. He has problems inside the U.S. delegation and in the White House, however, and we cannot now assess the international reaction nor his ability to cope with his problems. But they don't look as dark as they did two months ago" (Sam Levering 1981).

The resumed tenth session in Geneva (August 3–29) was a difficult one for all concerned. The U.S. delegation raised general concerns about Part XI, but it was instructed not to negotiate on specifics until the policy review was completed. The G77 wanted to move forward, but its leaders were reluctant to make concessions until the United States was ready to negotiate. Angry that Leigh Ratiner had opposed the negotiations during the Carter years and distrusting him personally, "his old cronies among the developing nations wouldn't talk to him" during the early weeks of the session, Lee Kimball recalled (Kimball 1995).

Our work was largely informal. Eleanor Smith recalled that she and others in the Neptune Group "interpreted the U.S. position to many delegates, especially the G77, to try to pacify their anger at the U.S. We feared that things could get out of hand" (Smith 1989). Lee, our resident expert on the draft treaty who remained a heavyweight at the conference, talked with scores of delegates about the U.S. objections and ways of moving the negotiations forward.

Our most useful contribution to the session may well have been *Neptune* number 18, prepared in advance largely by Lee and me and distributed to delegates at the start of the session. Rising to the challenge presented by the Reagan administration's policy review, Lee contributed several superb articles to this issue. In an editorial that set the tone, she argued that "new circumstances" called for "a reassessment of goals and tactics," and that the "new leadership" of the

---

7. Ted Kronmiller to James L. Malone, August 1981 (ML files). The State Department telegram, classified "confidential," contained the following note from Kronmiller to Malone: "The following is a memo from Sam Levering to myself. Although you had an opportunity to review it prior to your departure, it seems that you could find it useful at Geneva. As you are aware, Sam has an unusual rapport with the Group of 77 and his ideas should be considered in that light."

conference [read: Tommy Koh] needed to "grab the reins and preside." Another article pointed out that the only hope of preventing the treaty's defeat in the United States by those who opposed international institutions on ideological grounds was to make the seabed mining system "viable . . . in the eyes of the majority of those who would develop the seabed." Another article argued that joint ventures offered the best hope for achieving that objective (a point, she noted, that was first made in *Neptune* in 1975), and offered details about how to incorporate joint ventures into Part XI. Yet another piece offered suggestions to "streamline the machinery which has grown up around the seabed mining system" (*Neptune* 1981, 2,4,6)

Although at the time we were mostly critical of our government, in retrospect it seems clear that the G77 deserves some blame for failing to respond imaginatively and comprehensively to the Reagan challenge. Lee's suggestions in *Neptune* number 18 offered potentially useful components of such a response. Unfortunately, G77 negotiators followed the U.S. delegation's example and largely clung to the status quo at the August 1981 session; thus virtually no progress was made in meeting the seabed-mining concerns of the United States and other industrialized nations.

Toward a Disappointing Climax

Sam and I felt as if we were riding on an emotional roller coaster during the eight months between the end of the Geneva session in late August 1981 and the end of the final negotiating session in New York in late April 1982. Whenever it looked as if the administration might negotiate seriously and that a treaty including the United States was possible, we experienced emotional highs. At other times, when the antitreaty forces in Washington appeared to be winning, our spirits drooped. Caring deeply about the fate of the negotiations, we had to hope for the best; but knowing how many enemies of the treaty held high positions in the administration, we also had to fear the worst.

Our spirits rose and fell several times during the fall of 1981. September generally was upbeat. I learned on September 8 that Malone had found some flexibility in the G77's views at Geneva and hence was likely to urge the administration to continue with the negotiations (Miriam Levering 1981d). Even more gratifying was Secretary of State Haig's testimony before a congressional committee on September 17, in which he asserted that the United States "did intend to continue to participate fully in the Conference, with the clear objective of bringing it to a successful conclusion" (Schmidt 1989, 240).

Speeches by Kronmiller on September 30 and by Malone on October 19 were much less encouraging. Kronmiller told the American Mining Congress that "elements of the Draft Convention do not meet our national interests and objectives," and warned that parts of the text were "intended to promote a fundamental redistribution of wealth and power through measures congenial to the centrally-managed economies and through political decisionmaking dominated by the less developed countries" (statement 1981, 1–2).

In his speech Malone argued that, although improvements could be made in the treaty, it "is not certain yet whether these improvements taken as a package would be sufficient for our objectives and produce enough support for the Convention in the U.S." Malone also doubted that there would be enough time at the session the following spring to "restore sufficient balance to the Draft Convention" (Malone 1981, 4, 7).

Sam and I were especially upset by Kronmiller's remarks, which struck us as motivated more by ideological hatred of "centrally-managed economies" than by specific, practical objections to the draft convention. Indeed, we had concluded by this time that much of the Reagan administration's critique of the treaty stemmed from an ideological dislike of international institutions, Third World and Soviet bloc political and economic systems, and indeed almost any kind of government regulation of economic activity. From our perspective, a statement in Malone's speech epitomized the administration's ideological orientation: "We do not think that the American people want to see this nation drift toward global institutions which diminish our power and influence in world affairs" (Speech 1981, 2).

At the Public Advisory Committee meeting on November 3, Lee Kimball asked several sharp questions about the policy review and about why several important current documents (including the options papers being prepared for the president) were classified such that she and other nongovernmental members of the committee were not permitted to see them. Malone responded that he would seek to downgrade the classification on his recent congressional testimony, but made no promises in regard to the options papers (Public Advisory Committee 1981b, 35–37, 51–53).

The deeper issue Lee was raising was whether the NGO members of the advisory committee would be able to contribute directly to the shaping of U.S. policy, as they frequently had done under Richardson. Malone's and Kronmiller's answer, at least for liberal internationalists and environmentalists, basically was no. As Lee was leaving the meeting, a protreaty Defense Department official said to her, "Keep up the pressure."[8]

My spirits shrank when I heard from Sam after the meeting that, as I wrote a supporter the next day, "the forces out to frustrate Secretary Haig and even keep this country from returning to the negotiations at all are on the move to such a degree that strenuous action is required" (Miriam Levering 1981e). Within a week, however, I felt reassured that the administration was likely to decide to return. But I still questioned whether, as I wrote at the time, "the U.S. will have its proposals together in a coherent and realistic way" (Miriam Levering 1981f).

Seeking to influence the administration's thinking, Sam and I wrote letters in October to five prominent executives and attorneys associated with major U.S. shipping companies. We urged them to "weigh in" with key officials and to stress "the importance of the treaty and the necessity that the United States keep its list of desired changes reasonable." Because we did not know most of these people personally, we doubted that our letters would help our cause. In November I wrote a lengthy protreaty letter to someone we did know, White House aide Richard Darman, but again without apparent effect (Levering 1981g).

The efforts of our associate Bart Lewis of Philadelphia that fall may have been more successful. Lewis met several times with Robert McClements Jr., the president of the Sun Companies, a large oil company that belonged to one of the U.S. seabed mining consortia. Lewis won a limited victory: he succeeded in getting McClements to write President Reagan and "urge that the United States participate in the upcoming session" of the conference, but he failed in that McClements denounced the seabed mining provisions and praised the administration's efforts to "rectify a treaty which would work to the disadvantage of our country" (McClements 1981). Despite Lewis's best efforts, McClements had ended up sounding more like Ted Kronmiller than Elliot Richardson.

8. The quote from the Defense Department official is in a letter from Miriam Levering to Bill Fischer, November 4, 1981 (ML files).

Like McClements's letter, Reagan's announcement on January 29, 1982, that the United States would participate in the spring session left us caught in the middle of the emotional roller coaster. We were pleased that the administration would seek to negotiate a treaty; but we feared that, short of a miracle, the president's six objectives would not be achieved. The United States sought to guarantee access for mining companies and to remove production limitations, technology transfers, and other provisions that had taken years to negotiate (Schmidt 1989, 241).

Our fears seemed to be confirmed at a breakfast meeting that Lee Kimball organized for about eighty congressional staffers on February 20. The speaker was Alvaro de Soto of Peru, who had coordinated the G77 position on seabed mining at the conference for several years. He said that the conditions that the president's January 29th statement appeared to lay down "called into question every one of the fundamental elements of the 1970 Declaration of Principles adopted by the U.N. General Assembly as the basis for negotiations." He saw a willingness in the conference to examine possible improvements in the mining system, but not to undertake the major overhaul that the United States seemed to require. "This," he said, "could take five years to negotiate."[9] Painfully aware of what he called the "very big chasm" that existed between the position taken in Reagan's statement and the views expressed by G77 leaders such as de Soto, Tommy Koh commented in early March that it would be "a great challenge to narrow that [chasm] to something bridgeable" (*New York Times* 1982, 3A).

Steadfast in our commitment to achieving a treaty, the Neptune team (Lee Kimball, Arthur Paterson, Robert Cory, Barton Lewis, Barbara Weaver and Cecily Murphy of UMLSP, Sam, and myself) worked during the New York session (March 8–April 30, 1982) to bring the two sides together. Unfortunately, the problems went deeper than just bridging the gap between the two sides, for the G77 and the U.S. delegations were divided within themselves between moderates who wanted to make the compromises necessary to achieve a treaty, and hard-liners who were skeptical of making any concessions to the other side. To complicate matters further, the de facto head U.S. negotiator, Leigh Ratiner, a moderate, could not put his full

---

9. Alvara de Soto is quoted on page 58 of the chapter on Congress of Miriam Levering's unpublished 1984 manuscript (ML files).

energies into negotiating because he was drawn into repeated fights over his instructions between moderates and hard-liners in Washington. In the early days of the session, Washington hard-liners even prevented Ratiner from holding an official appointment and thus getting paid for his work!

What did we do in this final attempt to facilitate agreement at the conference? Three areas of activity stand out: first, our now familiar information role, which included an issue of *Neptune,* two press conferences, biweekly published summaries of developments at the conference, and many conversations with reporters; second, work to increase the effectiveness of the administration's moderates, notably Ratiner and Haig; and third, intense work to construct bridges between the various factions and viewpoints represented at the conference.

Of our many and diverse information activities, perhaps the most significant was an editorial Lee Kimball wrote for *Neptune* number 19 entitled "Taking the First Step." In it she argued that the seabed mining portions of the draft treaty, as currently written, might well be objectionable to future U.S. administrations as well as to the current one, and that it would be "very difficult" for the west Europeans and the Japanese to "leave their U.S. ally out in the cold" if the United States decided not to sign the treaty. Thus the G77 should entertain no illusions that the treaty would work well without U.S. participation. As she had done in her editorial in *Neptune* the previous August, Kimball concluded with a strong challenge to the conference:

> The greatest obstacle seems to be universal reluctance to take the first step toward re-drafting and re-ordering generally agreed portions of the text. While this will be a thankless task, it must be undertaken if a viable international law of the sea treaty is to exist. Better to reconstruct a workable system on the agreed foundations than to cling to unworkable structures and gut the system and thus the entire foundation. (*Neptune* 1982, 2)

Although all members of the Neptune Group can take pride in Lee's articles and indeed in *Neptune* generally, we may well have erred by expressing doubts about Leigh Ratiner in conversations with reporters, especially during the March–April 1981 session. Normally used as background without direct quotations from us, these criticisms—and others from people in the administration and on Capitol

Hill—found their way into articles in the *New York Times*, the *Washington Post*, and other newspapers.[10]

When Ratiner was subjected to stinging attacks in the press again as the New York session opened in March 1982, U.S. negotiator Bernard Oxman, a protreaty friend, spoke with Arthur Paterson and me and urged us to stop making critical comments about Ratiner to reporters. Oxman said that right-wing ideologues in Washington were trying to destroy both Ratiner and the treaty and that, because Ratiner represented the best hope for completing the treaty, we should not help them destroy him. Ratiner himself approached Paterson in the U.N. delegates lounge and told him that, if the personal attacks on him in the media did not stop, he would lose whatever standing he had both in Washington and at the Conference (Paterson 1989).

For the remainder of the 1982 session, Paterson and I portrayed Ratiner positively in conversations with journalists and delegates, and Sam used his friendship with Sen. Alan Cranston (Dem., Calif.) to persuade the liberal senator to stop criticizing him. To boost Ratiner, I wrote an op-ed piece for the *Christian Science Monitor* that depicted him as a "well-informed" negotiator who "now appears to represent the best hope of obtaining a treaty which includes the U.S." (Miriam Levering 1982). Having been influenced negatively by Sam's and my battles with Ratiner in the late 1970s and by the widespread distrust of him at the conference, I never thought that I would write a largely favorable article about him.

Meanwhile, Bart Lewis did what no one else in the Neptune Group had been able to do: he developed a working relationship with a high-level administration official, Secretary of State Haig. Lewis recalled how this relationship developed and what resulted from it:

> Haig had [as an assistant] a young man from Philadelphia, a lawyer.... I was able to enlist his support through some politicos here in Philadelphia who had given Haig support. So he was willing to listen, and I got to Haig through his young attorney. In fact, we even

---

10. For examples of stories that included sharp criticisms of Ratiner, see Bernard D. Nossiter, "Reagan's Delay on Sea Pact a Source of Dismay at U.N.," *New York Times*, March 5, 1981, 4; Don Shannon, "Reagan Seeks Total Review of Oceans Treaty," *Los Angeles Times*, March 10, 1981, 12; and Nicholas Burnett, "The Ex-Kennecott Lobbyist and the Scuttled Sea Law," *Washington Post*, March 22, 1981, 1D.

got a couple of audiences when things were getting towards the end of the law of the sea conference.

The big question was whether the U.S. would find a way around the problems with Part XI. We were running back and forth, talking to people in Washington and to Tommy Koh and so forth, to see if a formula could be developed that would be acceptable. In the end, it never seemed to work out. Mr. Haig was not willing to give it any public enthusiasm. (Lewis 1990)

As Lewis noted, he met with Tommy Koh in New York to tell him what he was learning through his contacts in Washington. Lewis recalled that Koh "made vivid how difficult it was to get the Third World countries to make major changes at that time just to suit the U.S. The whole treaty was a bunch of compromises anyhow, and they all locked together. You couldn't just isolate one item, because it was all part of a stack of cards that was put together as a unit" (Lewis 1990).

Just as Koh found it difficult to bring about major changes in the draft treaty in New York, so Haig faced determined enemies of compromise within the administration, including Edwin Meese, William Clark, Ted Kronmiller, and probably President Reagan himself. A close friend of Reagan's, Meese may well have had the greatest influence on this issue. We found no way to make contact with him, much less to soften his hard-line position. When Koh went to Washington near the end of the session to see if it made sense to extend the negotiations, he received no encouragement from anyone in the White House. Koh recalled his meeting with Haig during this trip: "I think Al Haig tried to turn the administration around. And he told me it [a recent meeting of U.S. officials on law of the sea] was a very rowdy session. There was blood all over the floor, his blood mainly. People—Ed Meese and others in the administration—just slaughtered him" (Koh 1990). No wonder Haig did not support the treaty publicly in April, as Lewis had hoped: the secretary of state clearly was losing the fight within the administration.

Our efforts to increase the effectiveness of Ratiner and Haig thus largely failed. It is true that Ratiner helped make some improvements in the draft treaty, and that in early April Haig briefly succeeded in making Ratiner's instructions more flexible (Schmidt 1989, 247). But both men lacked powerful friends in Washington: Kronmiller and other hard-liners restricted Ratiner's instructions during the crucial last two weeks of the session and then fired him when it ended, and an isolated Haig himself was fired a couple months later.

Our efforts to build bridges at the conference were more successful. Lee Kimball became a nerve center for many of the day-to-day negotiations because she was discreet and because many delegates trusted her judgment and her information. Lacking confidence in Ratiner and other U.S. negotiators, many delegates also counted on Lee to interpret to them just what the United States had to have on particular issues. My close relationship with Lee and with Tommy Koh also increased my stature, because delegates knew that I could provide them with accurate, up-to-date information, often obtained from Lee or from Koh.

During the session Lee and I frequently staked ourselves out in the delegates lounge at the UN. We sat a good distance apart, because we wanted delegates to be able to speak with us privately and to feel that they could learn the most intimate details of the negotiations. I was proud that Lee had become such an expert on the negotiations and that delegates with serious business now sought her out more frequently than they did me.

Lee and other Neptune Group members plus two officials of the UN secretariat worked up a comprehensive package of amendments that, several delegates believed in retrospect, more closely represented viable middle ground than other sets of compromises offered during the final session (Kimball 1995). Lee and I also worked closely with members of the Group of 11, the so-called "Good Samaritans," who presented compromise proposals that they hoped would be acceptable to the United States and to the G77. The G11 were developed nations, including Australia, Canada, the Netherlands, the Scandinavian countries, and Austria. Lee and I became especially close to Peter Bruckner of Denmark, who recalled that the Neptune Group had been "helpful in giving us a sense of where the median line [between the United States and the G77] might be" (Bruckner 1990).

Working as an informal liaison between Koh and Haig during the final days of the conference, I became convinced that the United States would accept the treaty if the G77 agreed to two additional changes, one in the clause that provided for the review of the treaty in the future and the other in the technology transfer articles. But G77 leaders, distrusting the Reagan administration, refused to make these changes. They told me that no matter what concessions they made to please the United States, the administration still would oppose the treaty because of its ideological approach. In short, they simply did not trust U.S. officials enough to make further concessions.

We kept working right up until the moment on April 30 when Koh, at the request of the U.S. delegation, called for a vote on the treaty. Not more than ten minutes before, we were talking with Secretary Haig's office in Washington, trying to find the one or two actions by the G77 that might permit the United States to vote for the treaty or at least to abstain. When I was on the floor moments before Koh entered to begin the voting, I asked delegate P. S. Jagota of India to help with one more try to make the treaty acceptable. He wanted one issue named, not several. "Technology transfer," I told him. But it was too late: the third Reagan shock—voting no after forcing the vote—was about to hit.

Although 130 nations voted for the Convention with only 4 against (the United States, Israel, Turkey, and Venezuela), the 17 nations that abstained included the Soviet Union and most west European and east European nations. Thus only three potential seabed mining states—Japan, France, and Canada—voted for the treaty, whereas three Western nations whose combined economic clout and seabed mining capability were far greater—the United States, Britain, and West Germany—either voted no or abstained. Regrettably, the negotiations that had been conducted on the basis of consensus did not end that way.

The statement issued jointly by OEP and UMLSP reflected our deep disappointment about the U.S. actions at the end of the conference. Urging the administration to sign the treaty despite having voted against it, we contended that the United States "should assess the full treaty package, bearing in mind the trade-offs required in a negotiation with over 150 national delegations of varied interests and ideology." We argued that "the Treaty is so essential to international comity that failure to sign it could deprive the U.S. of a seabed mining industry, [and] weaken the U.S. role in peacemaking, in global relationships, and institution building." Our feelings of anger and frustration came through most clearly in the following sentence: "We cannot afford to petulantly isolate our nation in today's world" (Statement 1982, 1–2). Try as we might to keep our spirits up as we left the UN that day, in truth we had hit bottom on the Reagan roller coaster.

Our Work Winds Down

After the excitement and ultimate disappointment of the last session, the events of the next several months seemed anticlimactic.

Aware that the administration's opposition to the treaty had grown even stronger after April 30, we experienced neither shock nor surprise when Reagan announced on July 9 that the United States would not sign the treaty at the signing ceremony in Jamaica in December. This decision disappointed us, but we were more upset by the decision, announced at the same time, not to exercise America's right, as a nonsignatory, to attend the Preparatory Commission meetings beginning in March 1983 as an observer.[11] As Lee Kimball commented at the Public Advisory Committee meeting on July 14, "it makes no sense to close off further participation in that process" (Public Advisory Committee 1982, 99).

Because most members of the Public Advisory Committee had been appointed under previous administrations, at the July 14 meeting Lee, Sam, and other internationalists on the committee got to enjoy one of their few victories of the Reagan years: a resolution to support U.S. participation in the Preparatory Commission passed by a vote of 18 to 5, with 3 abstentions (Public Advisory Committee 1982, 52). Although it lacked practical effect, this vote gave the internationalists a small measure of revenge for the administration's recent actions against the treaty. Lee and other Neptune Group members worked hard during the summer and fall to reverse the administration's decision on the Preparatory Commission, but without success.

One thing that kept us working even as the United States distanced itself from the treaty process was our belief that Reagan and other high officials were poorly informed as well as ideologically misguided on this issue. As Lee observed in an op-ed piece for the *Christian Science Monitor* on May 27, it seemed virtually certain that the treaty eventually would be ratified by the required sixty nations and thus enter into force as international law. Moreover, it seemed likely that several of the potential seabed mining states would participate in the treaty and secure their mining rights under it, thus undermining the current efforts of some U.S. and west European officials to develop a competing "mini-treaty" seabed mining regime. In the absence of a

---

11. Malone recalled that the administration decided against participating in the Preparatory Commission because its mandate was limited to implementing the treaty. Because the administration considered the treaty seriously flawed, it took the view that U.S. participation would be unproductive (Malone 1994).

widely recognized "mini-treaty" system, Kimball predicted that U.S. mining companies eventually would have to either abandon their plans or move their operations abroad. Either eventuality would shatter the Reagan administration's vision of a viable U.S. seabed mining industry operating outside the treaty system (Kimball 1982, 23).

For us, the administration's ignorance and simplemindedness on law of the sea was epitomized in something Reagan reportedly said during the National Security Council meeting of June 29 in which the decision was made not to sign the treaty. Reagan's alleged comment first appeared in a July 12 story by one of the *Washington Post*'s most respected reporters, Lou Cannon. "We're policed and patrolled on land, and there is so much regulation that I kind of thought that when you go out on the high seas you can do what you want," Reagan reportedly remarked. Cannon and other journalists cited this comment to suggest naive ignorance. During this discouraging time we needed no convincing.[12]

Developments in Washington between the summer of 1982 and the spring of 1983 were not all bleak. Having decided against the treaty, the administration moved toward declaring a two-hundred-mile exclusive economic zone, thus increasing U.S. authority beyond the control of fishing out to two hundred miles that had taken effect several years earlier. But the administration did not support the highly nationalistic legislation introduced in Congress by Congressman John Breaux and Sen. Ted Stevens that would have contradicted numerous provisions of the treaty. Led by the United Methodists, we fought this legislation on Capitol Hill. We were encouraged that an administration official asked Lee Kimball to look over a September 21 draft of the EEZ proclamation and let him know where it deviated from the treaty. And we were pleased—and pleasantly surprised—that the revised EEZ proclamation and the accompanying ocean policy statement, both issued by Reagan on March 10, 1983, were not nearly as contradictory to the treaty as they might have been.[13]

---

12. Lou Cannon, "Public Perception Seems Split in Its Perception of Reagan," *Washington Post*, July 12, 1982, 3. Liberal columnist Mary McGrory quoted Reagan's alleged comment to criticize him; she wrote that Reagan was "proud apparently that no data clogged his mind." McGrory, "Sailing the Sea Treaty Shoals Without the Lighthouse of Fact," *Washington Post*, July 22, 1982, 3.

13. For background to the EEZ proclamation, see Schmidt 1989, 261–64.

At signing session in Montego Bay in 1982. Bottom row, from left: Sam and Miriam Levering, Lee Kimball, Cecily Murphy, and Elliot Richardson. Top row: Barbara Weaver and John Temple Swing. Courtesy of Miriam Levering.

For Sam and me, the highlight of our work during the Reagan years was attending the signing ceremony at Montego Bay in Jamaica December 6–10, 1982. Also representing the Neptune Group were ten women with extensive involvement in the United Methodist Project: Kimball, Weaver, and Murphy from the staff and seven others who had provided unpaid leadership for the project. We wished that the ten to fifteen people who had made the greatest contributions to OEP and the U.S. Committee could have been there as well.

The fact that several of the Methodists had not met leaders of the conference and key U.S. negotiators provided a rationale for setting up appointments with conference leaders, including Bernardo Zuleta, Tommy Koh, Elliot Richardson, and the head of the U.S. delegation to the signing ceremony, Thomas Clingan. Each put his personal stamp on the same message: thank you for your help, and press on.

Clingan and Koh touched us most deeply. Clingan seemed genuinely hurt that he could sign only the largely meaningless final act, but not the convention. Even in his off-the-record conversations with us, however, he maintained his professionalism by not criticizing U.S.

policy. For his part, Koh began the proceedings with a speech that left us floating on cloud nine. After discussing eight reasons why he believed that the treaty would stand the test of time, he then detailed the factors that had led to what he saw as the unique and effective negotiating process at the conference. Included was a paragraph (quoted in full on page xviii) focusing on "the role played by the non-governmental organizations, such as the Neptune Group." All Neptune Group members in attendance were thrilled as Koh explained the "three valuable services" that our group had provided to the conference. We noted as well that he did not refer to any other NGO, though some, including the Sierra Club and other environmental groups that we had worked with, certainly could have been mentioned. Sam and I had gone to Jamaica partly to have our spirits lifted, and that had happened far beyond our expectations.

With the conference now officially over and with the Reagan administration's oceans policy largely set by March 1983, it was time for Sam and me to end our intensive work in Washington. That happened officially in May 1983, when at my request the board of directors of the Ocean Education Project decided to phase out OEP's Washington operations by the end of the year. The organization itself existed for several more years, but my continuing efforts to build support for the treaty in the United States came largely through my work as a member of the board of directors of Citizens for Ocean Law. Meanwhile, Sam returned to two longtime interests: improving U.S.-Soviet relations and lessening the risk of nuclear war.

Working for a comprehensive, widely accepted law of the sea treaty for more than a decade gave us experiences in helping build a more peaceful world that most peace activists (including us, before 1972) can only dream about. These experiences enriched our lives, and we are deeply grateful to all who gave time and money to further our common cause. Above all, we hope that our work will inspire others to devote themselves to the unending quest for a more just and peaceful world, a quest grounded in an enlightened vision of global community and an ever-expanding rule of law. If our experiences offer any guidance, that quest will be difficult—and worthwhile.

8

Assessing and Learning from the Neptune Group

[M]en and women can live as if their world was malleable in their grasp; and . . . to live in this belief is to be most authentically alive.
—Richard N. Goodwin, *Remembering America*

World law, like a coral reef or a manganese nodule, grows slowly by accretion.
—Miriam Levering

TWO TASKS REMAIN: assessing the Neptune Group's efforts from 1972 through 1982, and suggesting some lessons that other NGOs might wish to learn based on this experience. Drawing heavily on Miriam Levering's and Lee Kimball's writings and on interviews with Kimball and with other members of the Neptune Group and especially with officials with whom they worked, I begin by assessing the group's contributions at the conference. I then examine their work in the U.S. political system, which encompasses three interrelated arenas: the administration, Congress, and public opinion (including interest groups and the news media). I conclude by suggesting several lessons that may be useful to other NGOs working in national politics and international negotiations.

Assessing: The Group's Work at UNCLOS III

It seems indisputable that the Neptune Group made important contributions to the conference. Diplomats from the United States and other countries, UN officials, NGOs, and journalists attending the conference repeatedly stressed the Neptune Group's positive contributions. And so have leading scholars of the negotiations, including Clyde Sanger, Markus G. Schmidt, and James K. Sebenius.[1] Almost all of the fifty-eight U.S. and UN officials, diplomats, and NGOs whom Miriam Levering, Eleanor Smith, and I interviewed while preparing this book were in broad agreement with a 1982 comment by UN official Bernardo Zuleta: "[Tommy] Koh's statement [praising the Neptune Group's work] no doubt represents the views of the whole conference" (Zuleta 1982).

Besides those whom we interviewed, Zuleta's praise was echoed in a letter a young delegate from Mauritius, Anil Gayan, wrote to Miriam Levering in May 1981: "I would like on behalf of the delegation of Mauritius... to express my deepest thanks to the group you represent for helping me and other third world delegates to understand some of the basic issues concerning the subject matter of the Conference" (Gayan 1981). Perhaps because of her deep belief both in the unity and equality of all human beings and in the importance of the Neptune Group's work with delegates from developing nations, Miriam told me that she was more pleased and moved by this letter than by any other she received while working on law of the sea.

What were the Neptune Group's most important contributions to the conference? Lee Kimball summarized four of them in a letter to Koh in April 1980:

(1) bringing together constructively in a third-party [nonofficial] forum delegates, UN Secretariat members and individuals whose expertise can contribute to solving difficult LOS issues;
(2) serving as a third-party source of information through the seminars and *Neptune*;
(3) playing a communications facilitation role among delegates and the Secretariat through informal contacts, friendships and our receptions... and
(4) preparation of detailed background papers and reports related to the seminar program which are used by delegates and the Secretariat alike. (Kimball 1980c)

1. Sanger praised the Neptune Group's "outstanding work in both public education and in seminars" (Sanger 1987, 33). Clearly referring to the Neptune Group, Schmidt noted that "NGOs were helpful in facilitating dialogue between United States and key G77 negotiators on critical issues, or in getting it going again when it seemed to be deadlocked" (Schmidt 1989, 65). Sebenius described in detail the "important seminar... held under Quaker and Methodist auspices" at the summer session in New York in 1978 (Sebenius 1984, 31).

Kimball's four points were generally like the "three valuable services" that Tommy Koh described in his December 1982 speech (see Introduction). Both emphasized information and communication. Some of the information was provided, Koh noted, by "independent experts," and some was generated by members of the group through articles in *Neptune* or through the "background papers and reports" that Kimball usually wrote. While information was a crucial component of five of Koh's and Kimball's seven points (all but the third one in each case), communication was central to every one of them.

Although "information" and "communication" describe much of what the members of the Neptune Group did at the conference, these concepts by themselves fail to capture the essence of their contribution. A more helpful concept is what Kimball has variously called the "catalyst function" or "catalyst role," and which she and others also have identified as the "honest broker" role.[2] Kimball explained the catalyst role in a 1989 conference paper:

> [D]rawing on their independent status, NGOs may be able to promote or facilitate agreement . . . that is, they can act as a kind of "honest broker" to mediate compromises. Such meetings may also generate new ideas by bringing together individuals with varying specializations and backgrounds. This catalyst role is distinguished from the activist role . . . in that it contributes not to a particular substantive outcome but . . . to . . . reaching agreement or compromise *per se*. The wide range of NGO contacts contributes to this ability to facilitate agreement.[3]

In our interviews with them, officials and diplomats singled out as especially important various Neptune Group activities that, taken together, constituted the group's catalyst or honest broker role. Zuleta recalled that the group's "strongest suit was in bringing together delegates and people with real concerns [outside experts and others] to

---

2. Kimball developed the concept of the "catalyst function" in an unpublished conference paper from the late 1980s, "The Role of Non-Governmental Organizations in Antarctic Affairs" (ML files). Prof. Alan K. Henrikson of the Fletcher School of Law and Diplomacy has suggested yet another role for the Neptune Group at the conference: a "gadfly role," which I define for this NGO experience as working against a tendency among the delegates toward lethargy to move the negotiations forward (Henrikson 1995).

3. Lee A. Kimball, "The Role of Non-Governmental Organizations in the Implementation of the 1982 LOS Convention," 11–12, ML files. This paper, presented at a conference in the Netherlands in June 1989, was published in Soons 1990, 139–61.

exchange views and be frank without having to commit anyone... to take off coats and talk like human beings" (Zuleta 1982). Roy Lee of the UN Legal Office noted that the Neptune Group/MIT seminars "definitely accelerated negotiations on financial arrangements" (Lee 1982). U.S. delegate Bernard Oxman also emphasized the value of the seminars, seeing them as "both education [especially of Third World delegates] and the opportunity... to explore substantive issues without regard to national positions." (Oxman 1990)

Others stressed the group's role in facilitating informal communication and mutual respect between U.S. and Third World negotiators. Elliot Richardson recalled that "all the time they were helping to convey to me and George Aldrich and Tom Clingan and others in the U.S. delegation an awareness of the attitudes of the people [mainly Third World delegates] that the Leverings had been eating with and talking to" (Richardson 1990). To move the negotiations forward, U.S. delegate James K. Sebenius recalled, "Koh, Nandan, Evensen, Richardson, and others sought the help of NGOs—sometimes quietly, but there was no secret of the value" (Sebenius 1990). Another U.S. delegate, John Temple Swing, also insisted that the group played a crucial broker role in the negotiations:

> The greatest contribution in retrospect is facilitating understanding and trust of Third World delegates of the positions the United States was taking—had to take—and trust of us as human beings and individuals rather than representatives of the Great Satan. The Neptune Group provided a neutral setting and it was very useful... It was possible for many Third World delegates to see the U.S. negotiating team solving problems with them rather than imposing preconceived ideas because of Neptune's help. (Swing 1982)

Unlike these U.S. officials, Leigh Ratiner did not always work closely or comfortably with leaders of the Neptune Group. But he was just as emphatic about their contribution to the conference: "To the extent that I had success, I couldn't have done it without a Neptune Group. And I don't think any negotiator can. There's got to be a group of honest brokers off to one side.... There have got to be people that can get out messages that the negotiator... can't get out" (Ratiner 1989).

Although most scholars and diplomats have praised the Neptune Group's work at the conference, a couple U.S. officials whom we interviewed offered criticisms. Otho Eskin recalled that some of his colleagues in the State Department "were annoyed by the NGOs" and "thought they were a nuisance." He said that some officials "felt the Leverings were overly naive about the motives of Third World countries." Finally, he commented that "NGOs felt a treaty at any cost was desirable, but the U.S. government [except for some international lawyers] thought only a good treaty was desirable" (Eskin 1989).

Whereas Eskin suggested criticisms of the Neptune Group's ideas, George Aldrich, the head negotiator on seabed mining during most of the Carter administration, found fault with the group's actions. He argued that the group "should have pointed out that the mining industry had no clothes"—that is, that seabed mining would take place only in the distant future, if at all. He also raised important questions about the value of the numerous Neptune Group seminars on seabed mining: "I thought the brainstorming sessions were more helpful at the time than I did in retrospect. They gave more immediacy to the issues than they deserved. The more people who talked [at the seminars], the more they gave a sense it would work. . . . We didn't realize how unrealistic some of these things were. We were led to believe that the technology was much closer than it might be."[4]

4. Aldrich 1990. After reading this chapter, Lee Kimball responded to Aldrich's criticisms of the group:

> I don't quite get the Aldrich criticisms, because they attribute to the Neptune Group what were, in fact, widely held views by the experts at the time. We merely made these views available to the conference through expert resource people. That the mining industry had no clothes is something that the U.S. government might have picked up as well.
>
> More to the point, I don't think it was apparent to anyone in the late 1970s and early 1980s that the potential deepsea mining industry would become less economically viable; the assumption was that minerals prices would continue to rise and thus make it economic in the not too distant future, and this was reflected in the various projections by expert analysts of the minerals industry, within and outside the industry itself. Alternative land-based sources, and in particular substitution of other materials, notably synthetics, as well as conservation techniques that allowed lesser quantities to be used, significantly changed the economics in the 1980s and beyond.
>
> As to the technology, Aldrich may be correct that it was further away than thought; but again, the assumption was that research and development would intensify as the industry became commercially feasible, solving technological roadblocks. (Kimball 1995)

154 | CITIZEN ACTION FOR GLOBAL CHANGE

Of these criticisms, I find most valid the argument that the Neptune Group wanted "a treaty at any cost." The Leverings, Kimball, and other members of the Neptune Group fervently desired a treaty; I believe that this desire made them too willing to accept and defend provisions ("compromises") that, unless changed substantially by the Preparatory Commission, might well have blocked developed nations' access to seabed minerals.[5] Although some Neptune Group members may have been too sympathetic to Third World positions, they were not "overly naive" about the motives of the G77. On the contrary, they were deeply disturbed by the efforts of Third World "extremists" to gain complete control over the seabed mining system.

Aldrich's criticism has some merit—especially if one adopts the viewpoint that the Neptune Group was helping set up a parallel system for seabed mining (Enterprise plus mining companies sponsored by nations) that would never work in practice. Contemporary critics—including former U.S. negotiators Leigh Ratiner and Richard

---

5. Kimball responded to these judgments as follows:

The "treaty at any cost" is a fair criticism to a point. There may have been gradations of difference among members of the Neptune Group, but while we fervently supported a treaty, I don't believe on either Sam's or my part that we defended positions that would have blocked Western access to minerals. The need to conclude the treaty was driven by the process of negotiation, where the developing countries, having waited one year for the U.S. to get over its review and adjust its views, basically refused to wait any longer. The treaty was going to be adopted in 1982, even over U.S. objection. Nor were some of the basic compromises going to be gutted. Many of the industrialized nations felt equally strongly that the final compromises were sufficient response to U.S. concerns.

It is essential to note, however, that they combined this view with pragmatic and explicit recognition that further adjustments could be made in the Preparatory Commission [emphasis in original]. This was explicitly noted in speeches at the signing ceremony in Montego Bay, if not upon adoption [on] April 30, 1982. Some of us worked to ensure that the point was made by developing countries in their speeches.... The changes that were adopted in 1994 reflect changed circumstances and attitudes, and most of them emerged in discussions at the Preparatory Commission. UNCLOS may have concluded, but the process continued through 1994. (Kimball 1995)

Darman—were correct, I believe, in pointing out that most of the affected Western business interests and most members of the U.S. Senate were unlikely to support what was being negotiated on seabed mining in the late 1970s. But, as Aldrich and Richardson recognized at the time, trying to make the parallel system work was the only politically acceptable way at the conference to move the negotiations forward. Wanting a completed treaty, Aldrich, Richardson, and most other negotiators strongly supported the seminars when they were being held.

Agreeing with these negotiators' contemporary views, I believe that members of the group were correct to organize the seminars and thereby help to complete the treaty. Although they were too caught up in their desire to finish the negotiations to acknowledge how flawed Part XI was even after they had helped the negotiators make improvements in it between 1978 and 1980, they rightly foresaw that a completed treaty seriously flawed in its seabed mining provisions was preferable to a failed conference. The other parts of the treaty represented major steps forward in international law and governance, and should not be lost. They hoped that the United States and other industrialized nations would accept the seabed mining provisions as part of a compromise package deal. If not, perhaps Part XI could be modified sufficiently either in the Preparatory Commission or through additional negotiations after the ideological fires surrounding the conference had died down.

Those who supported continued U.S. participation in the negotiations, including the leaders of the Neptune Group and such former officials as Elliot Richardson, turned out to be prophetic. During the nearly twelve years between the close of the conference in December 1982 and the Clinton administration's acceptance of the revised treaty in July 1994, negotiations took place in a variety of forums to narrow the gap on seabed mining between the G77 position (as reflected in key sections of Part XI of the treaty) and the more probusiness position of the United States and several other major industrialized nations.

As supporters of the negotiations had hoped, the ideological fires surrounding the conference died down. The search for moderate solutions to the impasse on deep seabed mining accelerated after the collapse of communism in the former Soviet Union and eastern Europe and the glaring inefficiencies of state-managed enterprises in the Third World discredited statist models of economic development. The major changes in Part XI, accepted as part of the treaty by the United

Nations General Assembly in 1994, reflected a final victory for moderates such as Richardson and Koh whom the Neptune Group had sought to strengthen during UNCLOS III itself.

Assessing: The Group's Work in the United States

Unlike other NGOs who lobbied for particular positions, the Neptune Group's primary goal at the conference was to help build consensus on treaty issues rather than advocate a particular outcome. As we have seen, most delegates and UN officials appreciated this catalyst/honest broker role. In the U.S. political system, in contrast, the members of the group were advocates for the negotiations and opponents of seabed mining representatives and ideological conservatives, both inside and outside the government, who were skeptical that a treaty could be negotiated quickly or, assuming it was completed, that it would serve U.S. interests. Within the United States the members of the group thus were engaged in a decade-long political fight that left them with adversaries and defeats as well as allies and victories. In this fight their victories largely involved protecting the negotiations, while their defeats centered on failing to persuade the skeptical and the undecided that they should support the actual treaty being negotiated in the late 1970s and early 1980s.

As we have seen, the members of the Neptune Group labored long and hard to protect the negotiations. In their work with the administration, they helped strengthen treaty supporters (notably in State and Defense) and to weaken treaty opponents (notably in Treasury and Interior). Through their participation in meetings large and small and through the letters and phone calls to high administration officials and members of Congress that they generated, they showed that business representatives and environmental activists were not the only major nongovernmental influences on law of the sea.

Many officials also appreciated the group's publications and their persistence in pursuit of ocean law. "There is so much inertia in government," a Commerce Department official told Miriam Levering in early 1984. "You served, as someone must, as a catalyst."[6] But the group's influence on administration policy should not be exagger-

---

6. The unnamed official is quoted in Miriam Levering's "Administration" chapter in her 1984 manuscript, 68 (ML files).

ated: high officials debated and then finalized U.S. policy, a fact that largely helped the Neptune Group's cause from 1972 through 1980 and hurt it thereafter.

The Neptune Group also worked diligently to build media and public support for the negotiations. Over the years the two Leverings, Kimball, Arthur Paterson, Cecily Murphy, and other members of the group had countless conversations with U.S. and foreign journalists. The group also organized numerous well-attended press conferences at UNCLOS III, and appeared frequently on programs such as National Public Radio's "All Things Considered."

To build public support, the Methodist Project held conferences across the country and named regional coordinators of the protreaty effort within their denomination. Besides their considerable work with the Methodist Project and with other religious organizations, Kimball and the Leverings accepted invitations to speak to a wide variety of secular groups during these years. All of these efforts, however, failed to spark a broadly based national debate on law of the sea in general or on the merits of the "common heritage" principle in particular.[7]

In retrospect it is clear that the Neptune Group's most important work in U.S. domestic politics was with Congress. With Sam Levering setting the example of objective analysis of the ever-shifting balance of forces in Congress combined with dogged determination, members of the group mastered the legislative game. Among other things, they developed close ties with key protreaty lawmakers and their aides in both houses; they helped write bills and resolutions that countered the seabed mining bills and slowed the legislative process; they lined up protreaty figures to testify against the seabed mining bills; they alerted their supporters to write or call key representatives shortly before bills were to be voted on; and they called on allies in the administration to contact friendly legislators to insert provisions in committee that would make bills more congruent with the negotiations.

Without the Neptune Group's efforts, including in particular the Leverings' strong relationships with Rep. Berkley Bedell and

---

7. Arthur Paterson has noted that "over the ten years [1972- 82] there was little extended, meaningful debate in the American public over the common heritage" (Paterson, "Press" Chapter of unpublished manuscript, 74, ML files).

Ambassador Elliot Richardson, the seabed mining legislation almost certainly would have passed in 1978 or 1979. Without its efforts, the legislation that finally was enacted in 1980 might well have been more nationalistic than it was. The appreciation that Koh and other protreaty diplomats expressed for the Neptune Group's work on Capitol Hill to protect the negotiations from poorly timed, highly nationalistic legislation was not misplaced.

Although the members of the Neptune Group helped protect the negotiations in Congress, their influence on Capitol Hill declined as the conference dragged on. It was more persuasive to argue in 1973 or 1974 that the newly convening conference should be given time to reach a satisfactory conclusion before unilateral legislation should be passed than it was to make basically the same argument several years later when the conference appeared to be deadlocked on key issues. Similarly, the case for seeking comprehensive ocean law was easier to make in the early years, when the issue seemed to be either law congruent with U.S. interests or no law, than it was in later years when the choice was between this particular flawed treaty or possible alternatives outside it.

Ideas thus were more important than economic interests per se on this issue: during the early years the idea of advancing international law helped the Neptune Group and its allies defeat well-financed industry representatives on the seabed mining issue in Congress; but in later years more widespread, more persuasive criticisms of the emerging treaty tipped the scales toward the side of industry. In short, a poorly financed but determined NGO can do well in Congress, but only when its ideas command broad support.

Drawing on interviews with David Keeney of the staff of the Senate Foreign Relations Committee and others who worked on this issue on Capitol Hill, Markus G. Schmidt argues that, while NGOs (that is, the Neptune Group) were able to "alter the balance of power" in Congress "temporarily," by the late 1970s they were failing to make long-term gains for their cause:

> NGOs were looked upon by many Congressmen as "one-worlders" who were as myopic in their support of the LOS negotiations as their detractors were in their opposition. In particular, during the late 1970s, they strengthened the opinion of Representatives who already supported the conclusion of a Treaty but did not convert anyone who rejected the Committee One negotiating texts [on seabed mining] or the CHOM [common heritage of mankind] concept. The conclusion

cannot entirely be refuted that their support for it fueled conservative criticisms that the sea-bed regime, as it took shape, was a give-away to the Third World.[8]

I agree with Schmidt's first two points: that Neptune Group members working in Congress sometimes were perceived as "myopic" and that they did not "convert" their opponents (or more than a few undecided legislators) during the late 1970s or thereafter. But I believe that most of the fuel for "conservative criticisms" came from mining representatives and from nationalistic journalists and politicians, not from a backlash against protreaty NGOs. It also came from statements by Third World militants such as Paul Engo and from developments at the conference. Conservatives did not need fuel from NGOs to react in horror to the treaty's technology-transfer provisions and the prospective establishment of a heavily subsidized seabed-mining "Enterprise" that would compete with private firms.

Although some of our emphases differ, Schmidt and I agree that the Neptune Group's extensive efforts failed to stem the rising tide of skepticism on Capitol Hill about the merits of the treaty that Richardson and others were negotiating. But at least the Neptune Group had enough internationalist supporters in Congress in the early 1980s to beat back the efforts of Sen. John Breaux and other extreme nationalists to do even more than the Reagan administration was already doing to distance the United States from the ongoing process of creating a new international legal framework for the oceans.

Learning: Five Lessons

Finally, what lessons might other NGOs be able to learn from the Neptune Group's experience? Five stand out, the first of which is that NGOs need to decide early in negotiations whether they wish to be primarily insiders or outsiders.[9] If an NGO representative (for example,

---

8. Schmidt 1989, 67. "I always liked what they [members of the Neptune Group] did," David Keeney told us. "Some people may feel that you were too idealistic. Some people didn't agree with your position and felt that you were pressing too hard for the Third World position, and not for our [the U.S.] position. People identified you with the so-called 'internationalist' position" (Keeney 1989).

John Logue and Elisabeth Mann Borgese) early on develops a reputation as an outside critic, it is very hard later to shift gears and be accepted as an insider.

The members of the Neptune Group chose primarily to be insiders: that is, they decided to work directly—and discreetly—with powerful people, both in Washington and at the conference, and to seek to gain credibility and influence with these people. Sam Levering and Lee Kimball, for example, accepted official appointments to the Public Advisory Committee, though not to the U.S. delegation. The fact that the Neptune Group also played an outsider role in the U.S. domestic context by advocating ocean law and helping generate public and Congressional support for UNCLOS III strengthened the group's insider role, especially at the conference.

Several other NGOs chose largely to be outsiders: that is, they offered critiques of the negotiations and sought publicity for their views, but did not develop close relationships with leading officials, diplomats, and seabed mining representatives wary of their public advocacy. Because this study focuses on the Neptune Group, these lessons are directed to those who seek to increase their effectiveness as insiders in international negotiations.

A second lesson from the Neptune Group's experience is that, to establish an effective insider role, an NGO needs to develop what Lee Kimball calls "credibility" (Kimball 1989) and fellow group member Eleanor Smith calls "legitimacy" in the negotiations (Smith 1983, 40). Another word for what Kimball and Smith are describing is "acceptability."[10] Based on the Neptune Group's experience, I would define credibility/legitimacy/acceptability as confidence by many negotiators that an NGO can assist an international conference in

---

9. Lee Kimball has pointed out that both the insider and outsider roles can be effective: "There is a strong case to be made for both the 'insider' and the 'outsider' roles. Where the insider can participate in policy determinations bearing in mind the need to accommodate differing national and international positions, the outsider can more effectively generate public support by taking more uncompromising positions and make the 'insider' position seem more acceptable. Confrontation on the outside can increase responsiveness to stronger demands on the inside" (Kimball, "The Role of Non-Governmental Organizations in Antarctic Affairs," 50).

10. Prof. Alan K. Henrikson suggested "acceptability" as a synonym for "legitimacy" (Henrikson 1995).

reaching a broadly acceptable outcome. The major building blocks of the group's credibility/legitimacy/acceptability at the conference—the specific lessons that other NGOs most need to ponder—were these:

1. *Develop and share knowledge.* An NGO needs to develop and maintain a thorough, up-to-date knowledge of the issues being negotiated (including each issue's negotiating history) and a willingness to share information with others. Smith summarized the importance of knowledge for an international nongovernmental organization: "The real effectiveness of an INGO to the conference is in what they know that the delegates do not know, in what they can do that the delegates cannot, or in being able to present or explain information in a new or understandable light" (Smith 1983, 40).

2. *Separate facts from opinion.* If a delegate asks an NGO for facts about an issue, be careful not to answer with facts mixed with opinion. If a delegate wants opinion in addition to facts, make sure the opinion is clearly separate. "I try to be as clear as I can on what is fact, and what would be my opinion—or the organizations's opinion—of where you want things to go, and why," Lee Kimball recalled, and added that it is important not to try "to fool people by mixing up facts with your objectives" (Kimball 1989).

3. *Learn the process.* An NGO needs to develop a shrewd understanding of the personalities, the relationships, and the often informal ways in which decisions are reached in international conferences and in national governments. To disseminate information effectively, one must find out who listens to whom and what kinds of suggestions appeal to particular people (Kimball 1989). One must also be patient and understand the difficulties under which diplomats labor. Finally, one builds credibility/legitimacy by working with the negotiators over an extended period of time. In an interview with us, Kimball expressed concern that many NGO representatives in the contemporary environmental movement had not tried to grasp the process of international negotiations and thus tended to be less effective than they could be (Kimball 1989).

4. *Support the negotiators and avoid lecturing them or seeking credit for one's ideas.* "One of the first 'rules' mentioned to me by Miriam Levering," Smith observed, "was not to assert credit for any certain knowledge but to convey the information casually and let the ideas appear to be those of the delegate or member of the Secretariat. This way they have a greater chance of being used or at least supported or introduced into the negotiations" (Smith 1983, 53).

Miriam elaborated on basically the same lesson: "We learned not to preach to the delegates as if we had all the answers, to move quickly from the realm of what was most ideal to the realm of what was attainable in the conference, and to make it clear in all our relationships with the Secretariat and the delegates that we were working with them to find a solution, and were not outsiders telling them what to do" (Miriam Levering 1976b). In other words, humility, realism, and a genuine desire to help are cardinal virtues for NGOs.

5. *Advocate general advances in world order, not specific solutions.* Although it may well be counterproductive to push specific solutions, it is appropriate for NGOs to support such general goals as more comprehensive and widely accepted international law and increased protection for the environment. Indeed, because NGOs are not as buffeted as officials are by day-to-day pressures from diverse and often self-interested constituencies, they can offer clear and consistent support for long-term goals.[11] "More is at stake [in international conferences] than simply the protection of individual national interests," U.S. negotiator Bernard Oxman commented in 1979. "It is very important that there be groups [NGOs] which reflect that perspective, that there is a generalized concern about the international community, about international order that transcends the day to day concerns that we delegates worry about."[12]

Throughout the arduous, often unpredictable negotiations, the members of the Neptune Group held true to their belief that what Oxman called "international order" needed to be strengthened, and that UNCLOS III could make a major contribution to that process.

6. *Recognize that the lodestar for most diplomats is national interest.* It is easy for NGOs, as they rightfully enunciate long-term goals, to forget that diplomats are agents of nation-states whose primary job is to ensure that their governments' immediate interests are protected and advanced. Reflecting upon her many years of experience working on law of the sea and then on Antarctica, Lee Kimball offered this advice to NGOs: "The trick is to figure out how to use/harness

---

11. "In a sense," environmental lawyer Jim Barnes has written, "the international NGO phenomenon provides the cutting edge of the common interest" (Barnes 1984, 175).

12. Oxman was quoted in a publication of the Quaker United Nations Office in Geneva dated April 22, 1979 (ML files).

national self-interest to achieve the objectives you wish to promote; to couch preferred outcomes so that they are seen to serve national interests. If you can make good arguments, based on sound facts, that indicate how national self-interest is served by what you are recommending, then you get converts [among government officials and diplomats]" (Kimball 1995).

Although championing such ideals as broad international revenue sharing from seabed mining and advanced mechanisms for resolving disputes, the Neptune Group always recognized that the treaty would have to include numerous compromises that diplomats from around the world could use to persuade their superiors that their nations were winners in the negotiations. NGOs can be taken seriously when they encourage diplomats to consider an enlightened view of national interest, but not when they urge that specific national interests (for example, in resources within two hundred miles of land or in seabed mining) be sacrificed for an uncertain ideal.

The third broad lesson that emerges from the group's experience is the importance of building networks with other NGOs. As we have seen, the Neptune Group resulted from the close cooperation that developed in 1974 and 1975 between the two small Quaker/world-order groups headed by the Leverings and better-funded, longer-established United Methodist organizations in Washington. In turn, the Neptune Group organizations developed cooperative relationships on specific projects with at least a dozen other NGOs during the years of the conference. The most important of these, listed alphabetically, were the American Friends Service Committee, the Center for Environmental Law and Policy, the Connecticut Cetacean Society, the Friends Committee on National Legislation, the Humane Society of the United States, Members of Congress for Peace Through Law, the Methodist and Quaker programs in New York and the Quaker program in Geneva, the Sierra Club (including its then partly autonomous international office), the Stanley Foundation, the William Penn House, and the World Affairs Council of Philadelphia.

Besides working with the Neptune Group on specific projects, these and other NGOs often agreed to advocate the need for a new law of the sea—and to oppose unilateralism in Washington—in their publications. They thus built support for UNCLOS III among much larger and more diverse constituencies than the Neptune Group organizations could reach by themselves. Without this networking, the Neptune Group's achievements would have been more modest.

The fourth lesson is that people working in NGOs for worthy causes should make long-term commitments and not be discouraged by temporary setbacks. Lee Kimball and Miriam Levering are cases in point. Neither permitted President Reagan's 1982 decisions not to sign the treaty and not to allow U.S. diplomats to participate in the Preparatory Commission to dishearten them. Kimball remained very active throughout the 1980s as director of the Council on Ocean Law, the leading protreaty organization in Washington, and continued to work informally on treaty issues thereafter. Levering, a member of the board of directors of Council on Ocean Law until her death in November 1991, continued after 1982 to do everything in her power to prod foreign diplomats and U.S. and UN officials to make sufficient changes in the seabed-mining articles to enable the United States to sign the treaty.

The fifth and perhaps most important lesson is that, even in a world seemingly dominated by elected officials, by leaders of large public and private bureaucracies, and by others who attract attention from the media, ordinary people such as the Leverings and their associates—and midlevel diplomats such as Bernard Oxman and George Aldrich of the U.S. delegation—can contribute significantly to human betterment. They can do so, this study suggests, by having a vision of how the world can be made better—in this case, through the strengthening of international law and institutions—and then working diligently, persistently, and practically to further that vision. And they can be widely admired among those with whom they have worked even though, as in most human endeavors, their ideas and actions are open to criticism from some perspectives.

I thus conclude that, at least in this case, history was made more by determined, resourceful people—including people in particular NGOs and in particular governments and international institutions—than it was by the largely abstract structures and processes that contemporary social scientists emphasize as part of their often admirable effort to build "theory."[13] Although theory frequently has heuristic

---

13. Fairly typical in their general but varying emphasis on theory in studying international relations are the essays in Hermann et al. 1987. The emphasis on theory is even more pronounced in Keohane 1986. So strong is the emphasis on theory in much of the social science literature that Thomas Princen and Matthias Finger seem apologetic when they write that "a tight, logical theory of world environmental politics . . . is still a long way away" (Princen and Finger 1994, 14).

value, I believe that it is equally important to undertake detailed studies focusing on the role of human agency in history. For UNCLOS III, in my judgment, no theory will ever be rich enough to account for the positive, negative, and mixed contributions of, for example, Tommy Koh and Paul Engo, Elliot Richardson and Leigh Ratiner, Peter Bruckner and Alvaro de Soto, and Lee Kimball and Miriam Levering.

William Stelle, a counsel for the House Merchant Marine and Fisheries Committee, worked with many NGO and business representatives in Washington and followed the Neptune Group's work at UNCLOS III closely. With the Neptune Group's work in mind, Stelle reflected on the importance of individuals in NGOs in a 1985 letter to Kimball:

> It is my observation that the success of an NGO turns largely upon the character of the individuals who operate it. When the individuals have demonstrated proficiency in the subject matter, knowledge of the participants and experience in the process of negotiation, it is more than likely that their efforts to facilitate negotiation will be positively received, and will yield results. Suffice it to say that the converse is also true. Thus, the success of an NGO will also turn upon the strengths and abilities of the individuals who give it life. (Stelle 1985)

On balance, UNCLOS III (1973–82) and the subsequent negotiations leading up to the revised treaty's becoming international law twelve years later offer hope about the ability of fallible human beings, representing diverse and often conflicting constituencies, to work together on immensely complex problems. The law-of-the-sea saga suggests that the mistakes that people make, both individually and collectively, can themselves be rectified by individuals aware of their own potential power and determined to bring about change. On ocean management, global warming, nuclear proliferation, population growth, and myriad other contemporary issues, individuals building networks of support for their ideas—not abstract structures and processes—will largely determine whether the circumstances facing our human family and the natural environment that sustains us will improve or will deteriorate in the years ahead.

In June 1991, less than five months before her death, Miriam received a letter from a longtime supporter, a World Federalist from Minneapolis named Stanley Platt. "When the Treaty finally becomes effective, you and Sam will deserve much credit," Platt commented.

"Your long years of capable devotion will finally be rewarded, to the benefit of coming generations" (Platt 1991).

If Platt's comment is expanded to include Lee Kimball, Barton Lewis, Arthur Paterson, Sister Mary Beth Reissen, Barbara Weaver, and the scores of other dedicated people who worked in the Neptune Group and the hundreds more (including Platt) who made significant financial contributions to back their efforts, and if it is broadened to encompass the thousands of delegates, officials, business representatives, international lawyers, professors, journalists, and other NGO representatives who helped complete the treaty and build support for it around the world, his expanded statement becomes a fitting tribute to this landmark effort to strengthen ocean law and world order.

Appendix

References

Index

APPENDIX

Meetings of UNCLOS III

First session (organizational)	Dec. 3–14, 1973, New York
Second session	June 20 to Aug. 29, 1974, Caracas
Third session	Mar. 17 to May 9, 1975, Geneva
Fourth session	Mar. 15 to May 7, 1976, New York
Fifth session	Aug. 2 to Sept. 17, 1976, New York
Sixth session	May 23 to July 15, 1977, New York
Seventh session	Mar. 28 to May 19, 1978, Geneva, and Aug. 21 to Sept. 15, 1978, New York
Eighth session	19 Mar. to 27 Apr. 1979, Geneva, and July 19 to Aug. 24, 1979, New York
Ninth session	27 Feb. to 3 Apr. 1980, New York, and July 28 to Aug. 29, 1980, Geneva
Tenth session	Mar. 9 to Apr. 24, 1980, New York, and Aug. 3–9, Geneva
Eleventh session	8 Mar. to 30 Apr. 1982, New York
Twelfth session (signing)	Dec. 6–10, 1982, Montego Bay

References

Abourezk, James. 1983. Letter to Miriam Levering, May 17. ML files.
Aldrich, George (U.S. official). 1990. Interview. March 27. ML files.
Baratta, Joseph Preston. 1982. "Bygone 'One World': The Origin and Opportunity of the World Government Movement, 1937–1947." Ph.D. diss., Boston Univ.
———, comp. 1987. *Strengthening the United Nations; A Bibliography on U.N. Reform and World Federalism.* Westport, Conn.: Greenwood Press.
Barnes, James N. 1984. "Non-Governmental Organizations: Increasing the Global Perspective." *Marine Policy* 8: 171–81.
Bedell, Berkley (member of Congress). 1990. Interview. April 30. ML files.
Bleicher, Samuel (U.S. official). 1990. Interview. March 29. ML files.
Bridgman, Jim. N.d. Diary and notes relating to UNCLOS III. ML files.
Bruckner, Peter (Danish diplomat). 1990. Interview. April 4. ML files.
Chatfield, Charles. 1992. *The American Peace Movement: Ideals and Activism.* Boston: Twayne.
Church, Frank, and Jacob K. Javits. 1979. Letter to Cyrus R. Vance. October 30. ML files.
Citizens for Ocean Law. Undated brochure (1981?). ML files.
Curtis, Clifton (environmental activist). 1990. Interview. April 5. ML files.
DeBenedetti, Charles. 1980. *The Peace Reform in American History.* Bloomington: Indiana Univ. Press.
Dewar, Helen. 1979. "Would-Be Space Colonists Lead Fight Against Moon Treaty." *Washington Post*, October 30, 3A.
Diamante, John. 1978. "Neptune Plaudits from Geneva U.N. Press." Unpublished paper, May 23. ML files.
Dubs, Marne (mining industry representative). 1990. Interview. April 3. ML files.
Eskin, Otho (U.S. official). 1989. Interview. December 6. ML files.
Feraru, Anne Thompson. 1974. "Transnational Political Interests and the Global Environment." *International Organization* 28 (Winter), 31–60.

Forkan, Patricia (environmental activist). 1989. Interview. December 6. ML files.
Friedman, Wolfgang. 1971. *The Future of the Oceans.* New York: George Braziller.
Gayan, Anil. 1981. Letter to Miriam Levering, May 11. ML files.
Gayley, Margaret (congressional assistant). 1990. Interview. June 27. ML files.
Glassner, Martin (professor, Neptune Group associate). 1990. Interview. March 30. ML files.
Godwin, Richard N. 1988. *Remembering America; A Voice from the 1960s.* New York: Harper and Row.
Hamlin, Joyce (Neptune Group member). 1989. Interview. December 7. ML files.
Hearings Before the Subcommittee on Arms Control, Oceans and International Environment of the Committee on Foreign Relations of the United States. 1978. Ninety-fifth Congress, second session on S.2053. August 17.
Hearings Before the Subcommittee on Oceanography and the Committee on Merchant Marine and Fisheries. 1977. Serial No. 95-4.
Henrikson, Alan K. 1995. Letter to Ralph Levering, February 24. ML files.
———, ed. 1986. *Negotiating World Order: The Artisanship and Architecture of Global Diplomacy.* Wilmington, Del.: Scholarly Resources.
Hermann, Charles F., Charles W. Kegley Jr., and James N. Rosenau, eds. 1986. *New Directions in the Study of Foreign Policy.* Boston: Allen and Unwin.
Invitation to Ecumenical Prayer Service for the Law of the Sea Negotiation. May 23, 1977. ML files.
Irwin, Robert. 1982. "Anita Yurchyshyn and the Law of the Sea." *Sierra* 67: 142–44.
Katz, Ronald. 1979. "Financial Arrangements for Seabed Mining Companies: An NIEO Case Study." *Journal of World Trade Law* 13: 202–22.
Keeney, David (congressional assistant). 1989. Interview. December 4. ML files.
Kent, George. 1975. "Ventures and Misadventures." *Neptune* (May 7), 3, 9.
Keohane, Robert O., ed. 1986. *Neorealism and Its Critics.* New York: Columbia Univ. Press.
Keohane, Robert O., and Joseph S. Nye Jr. 1977. *Power and Interdependence: World Politics in Transition.* Boston: Little, Brown.
———. 1987. *Power and Interdependence.* 2d ed. Glenview, Ill.: Scott, Foresman.
Keohane, Robert O., and Joseph S. Nye Jr., eds. 1972. *Transnational Relations and World Politics.* Cambridge: Harvard Univ. Press.
Kimball, Lee. 1975a. Letter to Cyril Ritchie. n.d. (January?). ML files.
———. 1975b. Letter to Brad and Hannah Rishel, January 17. ML files.

———. 1978. Letter to Sam and Miriam Levering, May 10. ML files.
———. 1979a. "Package." *Neptune* 14 (April), 3.
———. 1979b. Letter to Cyrus Vance, November 8. ML files.
———. 1980a. Letter to Miriam Levering, November 18. ML files.
———. 1980b. "The Enterprise: Horse, Camel, or Duck-Billed Platypus." *Neptune* 16 (March), 1, 3.
———. 1980c. Letter to Tommy Koh, April 11. ML files.
———. 1981a. "Editorial." *Soundings* 6:1 (April–May), 5.
———. 1981b. Letter to James L. Malone, June 22. ML files.
———. 1982. "Taking the First Step." *Neptune* 19 (March), 2.
———. 1982. "After That 'Victory' on Law of the Sea." *Christian Science Monitor* (May 27), 3.
———. "The Role of Non-Governmental Organizations in Antarctic Affairs." N.d. (1987?) ML files.
——— (Neptune Group member). 1989. Interview. October 3.
———. 1990. "The Role of Non-Governmental Organizations in the Implementation of the 1982 LOS Convention." In *Implementation of the Law of the Sea Convention Through International Institutions,* edited by Alfred H. A. Soons, 139–61. Noordwijk aan Zee, Netherlands: Law of the Sea Institute.
———. 1995. Letter to Ralph Levering, January 4. ML files.
Kitsos, Tom (congressional assistant). 1990. Interview. February 28. ML files.
Koh, Tommy T. B. 1982. "Statement by President of Law of Sea Conference at Opening Meeting at Montego Bay Session." United Nations press release, December 6. ML files.
——— (diplomat from Singapore). 1990. Interview. April 11. ML files.
Koo, Samuel. 1978. "Clash over Sea Mining Likely at Conference." *Philadelphia Inquirer* (August 21), 8A.
Krueger, Robert. 1978. Letter to Richard Maxwell, May 19. In Robert Krueger papers, box 58, 1978 folder, library of the Univ. of Virginia School of Law.
Lacey, John Edward. 1982. "Role of Non-Governmental Organizations in the Third United Nations Law of the Sea." Ph.D. diss., Rutgers Univ.
Leach, Jim (member of Congress). 1990. Interview. March 29. ML files.
Lee, Roy (UN official). 1982. Interview. December 15. ML files.
Levering, Miriam. 1975a. Letter to William Fischer, December 16. ML files.
———. 1975b. Notes on talk by Louis Sohn at William Penn House in Washington, November 7. ML files.
———. 1976a. Letter to Maxwell Stanley, June 1. ML files.
———. 1976b. Letter to Phyllis D. Macey, March 6. ML files.
———. 1977a. Letter to Barton Lewis and William Fischer, June 20. ML files.

———. 1977b. Letter to Mary Lane Hiatt, April 18. ML files.
———. 1977c. Letter to Roderick Ogley, September 9. ML files.
———. 1978a. Letter to Barton Lewis and Bill Fischer, February 8. ML files.
———. 1978b. Letter to Barton Lewis and Bill Fischer, September 1. ML files.
———. 1979. Letter to Jim Magee, September 12. ML files.
———. 1980a. Letter to McNeill and Louise Smith, February 8. ML files.
———. 1980b. Letter to Elliot Richardson, May 7. ML files.
———. 1981a. Letter to Friends [Quaker] World Committee for Consultation, March 4. ML files.
———. 1981b. Letter to Eleanor Smith, March 18. ML files.
———. 1981c. Letter to Beatrice Howell, April 6. ML files.
———. 1981d. Letter to Robert L. Stuart, September 8. ML files.
———. 1981e. Letter to Barbara Hollingsworth, November 4. ML files.
———. 1981f. Letter to Bill Fischer, November 13. ML files.
———. 1981g. Letter to Richard Darman, November 11. ML files.
———. 1982. "Law of the Sea's Tempest-tossed Negotiator." *Christian Science Monitor.* (March 31), 2A.
Levering, Sam. 1977. Notebook for 1977. ML files.
———. 1978. Letter to editor. *Washington Post,* February 24, 20A.
———. 1979. Letter to John and Katherine Paterson, December 5. ML files.
———. 1980. Letter to Donald L. Guertin, July 19. ML files.
———. 1981. Letter to Stanley K. Platt, August 2. ML files.
——— (Neptune Group member). 1990. Interview. March 10. ML files.
Levering, Sam, and Miriam Levering. 1979. "Seabed Mining." *Neptune* 14 (April), 1,4.
Lewis, Barton (Neptune Group member). 1990. Interview. March 29. ML files.
Los Angeles Times. 1980. "Sea Law Session Ends; Gains Hailed" (August 30), 12A.
Magee, Jim. 1980a. Memorandum from Jim Magee to Miriam Levering, February 5. ML files.
———. 1980b. Letter to Miriam Levering, May 11. ML files.
Malone, James L., 1981. Speech at the American Enterprise Institute, October 19. ML files.
——— (U.S. official). 1994. Interview. July 18. ML files.
McClements, Robert Jr. 1981. Letter to President Ronald Reagan, November 12. ML files.
McColl, Richard (congressional assistant). 1990. Interview. March 29. ML files.
McLean, John (Neptune Group member). 1989. Interview. October 4. ML files.
Miles, Edward (professor, consultant to UNCLOS III). 1990. Interview. February 26. ML files.

Mott, Jessica (Neptune Group member). 1989. Interview. October 4. ML files.
Mount, Lucia. 1979. "Law of the Sea Parley Seeks Fair Sharing of Wealth." *Christian Science Monitor.* (March 19), 9.
New York Times. 1976a. "Compromise at Sea?" (August 21), 20A.
———. 1976b. "The U.N. at Sea" (September 22), 40A.
———. 1979. "U.S. Delegate Reports Progress at Sea-Law Session" (April 19), 20A.
Neptune. 1981. "Reassess and Act." No. 18 (August), 2.
Nordquist, Myron (U.S. official). 1989. Interview. December 7. ML files.
Nossiter, Bernard D. 1980. "Treaty on Moon: Is It Too Soon?" *New York Times* (March 9), 8E.
———. 1982. "U.S. Is Returning to Sea Treaty Talks." *New York Times* (March 5), 3A.
Ogley, Roderick. 1975. "New USSR Proposal; ISRA Could Mine Seabed." *Neptune* (April 4), 1, 8.
Orr, Jim (Neptune Group member). 1989. Interview. December 6. ML files.
Oxman, Bernard (U.S. official). 1990. Interview. February 26. ML files.
Paterson, Arthur. 1984. Unpublished paper on the Neptune Group's dealings with the media. ML files.
——— (Neptune Group member). 1989. Interview. October 3. ML files.
———. 1995. Letter to Ralph Levering. ML files.
Pelham, Ann. 1978. "Seabed Mining Bill Awaiting Action by Senate Committee." *Congressional Quarterly* (August 5), 2073.
Percy, Charles H. 1978. Letter to Barbara Weaver and Sam Levering, September 25. ML files.
Platt, Stanley K. 1991. Letter to Miriam Levering, June 21. ML files.
Pontecorvo, Guilio, ed. 1986. *The New Order of the Oceans.* New York: Columbia Univ. Press.
Princen, Thomas, and Matthias Finger. 1994. *Environmental NGOs in World Politics.* New York: Routledge.
Public Advisory Committee, U.S. Department of State. 1978. Transcript of March 2 meeting. ML files.
———. 1979a. Transcript of March 8 meeting. ML files.
———. 1979b. Transcript of May 18 meeting. ML files.
———. 1980. Transcript of November 13 meeting. ML files.
———. 1981a. Transcript of June 8 meeting. ML files.
———. 1981b. Transcript of November 3 meeting. ML files.
———. 1982. Transcript of July 14 meeting. ML files.
Raiffa, Howard. 1982. *The Art and Science of Negotiation.* Cambridge: Belknap Press.
Ratiner, Leigh (U.S. official, lobbyist). 1989. Interview. December 5. ML files.
"Report on the Conference on Law of the Sea: Financial Arrangements, A." 1978 (January 28–29). ML files.

"Report of the Conference on the International Seabed Enterprise and Authority, A." 1977 (February 4–6). ML files.
Richardson, Elliot (U.S. official). 1990. Interview. April 4. ML files.
Samuelson, Robert J. 1977. "Law of the Sea Treaty—Talk, Talk, and More Talk." *National Journal* (August 27), 1343.
Sanger, Clyde. 1987. *Ordering the Oceans: The Making of the Law of the Sea.* Toronto: Univ. of Toronto Press.
Scharlin-Rambach, Patricia (environmental activist). 1989. Interview. December 6. ML files.
Schmidt, Markus G. 1989. *Common Heritage or Common Burden? The United States Position on the Development of a Regime for Deep Sea-bed Mining in the Law of the Sea Convention.* New York: Oxford Univ. Press.
Sea Breezes. 1977. "First Quarter Touchdowns." 4:6 (November), 1.
Sebenius, James K. 1981. "The Computer as Mediator: Law of the Sea and Beyond." *Journal of Policy Analysis and Management* 1: 77–95.
———. 1984. *Negotiating the Law of the Sea: Lessons in the Art and Science of Reaching Agreement.* Cambridge: Harvard Univ. Press.
"Seminar on the M.I.T. Model on the Profitability of Deep Seabed Mining, August 29, 1978." Unpublished paper, ML files.
Short, Ken. 1980. "The U.S. Senate and the Law of the Sea Treaty: Some Prospects for Ratification." Unpublished paper, ML files.
Smith, David N., and Louis T. Wells Jr. 1975. *Negotiating Third World Mineral Agreements.* Cambridge: Ballinger.
Smith, Eleanor. 1983. Untitled paper on NGOs at UNCLOS III. ML files.
———. (Neptune Group member). 1989. Interview. October 5. ML files.
Smith, Jackie, Charles Chatfield, and Ron Pagnucco. 1997. *Transnational Social Movements and Global Politics; Solidarity Beyond the State.* Syracuse: Syracuse Univ. Press.
Smith, Jackie, Ron Pagnucco, and Winnie Romeril. 1993. *Transnational Social Organizations in the Global Political Arena.* South Bend: Joan B. Kroc Institute for International Peace Studies, Univ. of Notre Dame.
Soundings. 1976. "The Law of the Sea Conference." 2:1 (October), 1.
"Statement by Elliot L. Richardson," September 15, 1978. ML files.
"Statement of Theodore G. Kronmiller... Before the American Mining Congress Mining Convention, Denver, Colorado, 30 September 1981." ML files.
Statement, April 30, 1982, by OEP and UMLSP. ML files.
Stelle, William. 1985. Letter to Lee Kimball, February 11. ML files.
Stevenson, John R. (U.S. official). 1990. Interview. April 5. ML files.
"Summary of the Remarks of the Speakers from Five Panel Discussions on UNCLOS III, Held at Quaker House, Geneva, 11 April to 9 May 1978." ML files.
"Summary Report of the Three Seminars on Saturday, July 21, 1979." ML files.

Swing, John Temple (world order activist and U.S. official). 1982. Interview. December 15. ML files.
Taft, George (U.S. official). 1990. Interview. February 27. ML files.
Ten Commandments of the New Earth. 1979. UMLSP. ML files.
Tilly, Charles. 1984. "Social Movements and National Politics." In *Statemaking and Social Movements: Essays in Theory and History,* edited by Charles Bright and Susan Harding, 297–317. Ann Arbor: Univ. of Michigan Press.
Todd, James E. 1973. "The Role of NGOs at United Nations Conference on the Human Environment." *International Associations* (January), 42–45.
United Technologies. 1980. "Stranglehold on the Moon." *Washington Post* (February 14), 2A. ML files.
Wadlow, Rene. 1975. Letter to Lee Kimball, January 21. ML files.
Washington Post. 1977. "Sea Treaty at Sea" (July 25), 20.
Weiss, Thomas G., and Robert S. Jordan. 1975. "The Role of NGOs in the World Food Conference." *International Associations.* (January), 42–45.
Weiss, Thomas G., and Leon Gordenker, eds. 1996. *NGOs, the UN, and Global Governance.* Boulder: Lynne Rienner.
Whales: How Safe the Gentle Giants of the Sea? 1978. UMLSP. ML files.
Willetts, Peter, ed. 1982. *Pressure Groups in the Global System: The Transnational Relations of Issue-Oriented Non-Governmental Organizations.* New York: St. Martin's.
Wilson, George C. 1973. "Battle Stirs over Seabed Mines Bill." *Washington Post* (May 6), 1A.
Wooley, Wesley T. 1988. *Alternatives to Anarchy: American Supranationalism since World War II.* Bloomington: Indiana Univ. Press.
Yurchyshyn, Anita (environmental activist). 1990. Interview. April 2.
Zuleta, Bernardo. 1978. Letter to Miriam Levering, January 30. ML files.
——— (UN official). N.d. (December 1982?). Interview.

Index

Abourezk, James, 119
Adjali, Mia, 26
AFL-CIO, 37
Africa, 60
Aldrich, George, 152, 164; on environmentalists at Geneva, 80–81; on Sam Levering, 76–77; and Neptune Group, 129, 152–55; opens up U.S. policymaking process, 76; Reagan administration reassigns, 128; works with Congress in 1978, 115
Amerasinghe, Hamilton, 91
American Association of University Women, 41
American Baptist Church, 24
American Freedom Association, 11, 13, 24. *See also* Levering, Miriam; Ocean Education Project
American Friends Service Committee, 8, 65, 69, 163
American Mining Congress, 23, 37, 38, 137
Andrew, Dale, 58, 61
Antrim, Lance, 91, 93
Arlie House, 8
Associated Press, 82. *See also* news media
Australia, 86, 143
Austria, 143

Bailey, John, 86
Baptist Church. *See* American Baptist Church
Barnes, Jim, 32, 79, 80
Barstow, Robbins, 79, 81
Bedell, Berkley, 39, 110, 114, 119, 157
Bedell, Eleinor, 110
Bleicher, Samuel, 74
Boettcher, Robert, 40, 41–42
Borgese, Elisabeth Mann, 33, 61, 160
Brazil, 43
Breaux, John, 109, 112, 113, 119, 129, 146, 159
Brewer, William, 98–99, 122
Bridgman, Jim, 25, 52, 58, 59, 60, 67, 87
Brockett, Ted, 98
Brown, Jim, 69
Bruckner, Peter, 143, 165
Bryn Mawr College, 84
Buchanan, Patrick J., 107
Business Week, 69
Byrd, Robert, 119

Canada, 22, 39, 143, 144
Carnegie Endowment for International Peace, 28, 56
Carroll, Isabel, 103

Carter, Jimmy, 50, 74
Carter administration, 72–73, 74, 103–105, 107, 118, 121. *See also* Aldrich, George; Carter, Jimmy; Richardson, Elliot; Vance, Cyrus
Case, Clifford, 39
Catholics. *See* Center for Concern; Riessen, Mary Beth; Roman Catholic Church
Center for Concern, 41, 42
Center for Environmental Law and Policy, 163
Center for Law and Social Policy, 32, 79
Center for Ocean Law and Policy, 111
Central Intelligence Agency, 14, 60
Chase Manhattan Bank, 86
China, 67
Christian Science Monitor, 69, 120, 141, 145–46
Chudson, Walter, 86
Church, Frank, 123–24
Citizens for Ocean Law, 125, 148
civil rights movement (U.S.), 9
Clark, William, 128, 132, 142
Clingan, Thomas, 147–48, 152
Clusen, Charles, 38
"cod wars," 3, 43
Colman, C. J., 67
Commerce Department, 44, 91, 98, 122, 156
communist nations, 59–60. *See also* China; Cuba; Union of Soviet Socialist Republics; Vietnam
Conant, Melvin, 125
Congress, 76, 96; assessment of work with Neptune Group, 157–59; establishes Public Advisory Committee, 47; and fishing legislation, 43–46; on Sam Levering, 12; and Neptune Group during Reagan years, 131, 146; and seabed mining legislation, xix, 20– 21, 29, 36–42, 106–22, 126–27. *See also* Levering, Sam; seabed mining; United States Committee for the Oceans

Congressional Research Service, 40–42
Connecticut Cetacean Society, 79, 163
Corbett, J. Elliot, 42
Cory, Robert, 20, 24, 139
Council on Foreign Relations, 125
Council on Ocean Law, 164
Cousteau, Jacques, 61-62
Cranston, Alan, 39, 42, 45, 46, 141
Cuba, 107

Darman, Richard, 76, 117, 138, 154–55
Deepsea Ventures, 49
Defense Department, 4, 20, 42, 48, 131, 138, 156
Dellums, Ron, 114
Demars, Ken, 67
Denmark, 52, 143
de Soto, Alvaro, 92, 139, 165
developed nations, 3, 4, 61, 64, 73
developing nations: during Carter presidency, 85, 92, 118; during Ford presidency, 46, 59–60, 65–66; during Reagan presidency, 134, 139, 143; views of delegates from, xix, 3, 21
Diamante, John, 58–59
Dingell, Charles, 45
Downing, Thomas, 37
Dubs, Marne, 12, 78, 109, 110, 111

Egypt, 85
Eisenstadt, Stuart, 74
El Hussein, Ali, 102
Ely, Northcutt, 109, 111–12, 114
Engo, Paul, 66, 67, 87, 159, 165; heads negotiating group 3 (1978), 89; and M.I.T./Harvard seabed mining model, 94; and 1977 action, 108; and pro-Third World seabed mining articles (1977), 76; and seabed mining text (1975), 64–65; on Anita Yurchyshyn's environmental suggestions, 80

environmental issues and organizations, 38, 165; accomplishments at UNCLOS III for, 31–32, 78–81; and environmental provisions of seabed mining bills, 120; lessons of UNCLOS III for, 161; and Sam Levering, xviii, 75; at 1975 Geneva session, 62; and Sierra Club-Ocean Education Project program at Caracas (1974), 51, 53; in the work of the United Methodist Project, 10. *See also* Center for Environmental Law and Policy; Connecticut Cetacean Society; Cousteau, Jacques; Humane Society of the United States; Oceanic Society; Sierra Club
environmental movement (U.S.), 9
Equador, 45
Eskin, Otho, 29, 109–10, 152–53
Evensen, Jens, 76, 152

Federation of American Scientists, 41
Fenwick, Millicent, 116–17
Ferguson, Robert, 126
Fiji, 104
Fischer, William Jr., 9, 19–20, 45, 49
fishing issues, 43–46
Flipse, John, 49
Ford, Gerald, 45, 49
Ford administration, 49, 73
Forkan, Patricia, 32, 79, 80, 81
France, 144
Fraser, Donald, 39–40, 41–42, 74, 111, 112, 113
Friedmann, Wolfgang, 39
Friends Committee on National Legislation, 11, 19, 74, 163. *See also* Quakers
Friends United Meeting, 11, 13. *See also* Quakers
Friends World Committee for Consultation, 18. *See also* Quakers
Fye, Paul, 125

Gayan, Anil, 150
Geneva Convention on Fishing on the High Seas (1958), 44
Germany, Federal Republic of, 60, 118, 144
Glassner, Martin, xvii, xviii
Glomar Explorer, 22
Goldberg, Arthur J., 23
Gonzales de Leon, Antonio, 85
Graham, Frank Porter, 39
Gravel, Mike, 45
Great Britain, 3, 43, 144
Greenwald, Richard, 111
Griffin, Marvin, 45
Group of 11 developed nations, 143. *See also* developed nations
Group of 77 developing nations, 64, 141. *See also* developing nations
Gude, Gilbert, 14
Guyana, 43

Hahn, Lili, 24
Haig, Alexander, 128, 132, 136–37, 138, 141–42, 143
Hall, David, 52
Hamlin, Joyce, 14, 26, 56, 59, 60
Harvard Business School, 86. *See also* MIT/Harvard econometric model
Harvard Law School, 85
Heyerdahl, Thor, 53, 55
Hilmy, Joseph, 67, 85
Hollick, Ann, 58
House Foreign Affairs Committee, 121. *See also* House International Relations Committee
House Interior Committee, 114
House International Relations Committee, 74, 114, 115. *See also* House Foreign Affairs Committee
Houseman, Charles, 86
House Merchant Marine and Fisheries Committee, 41, 44, 45, 109, 113, 114, 117, 165
Hudson, Richard, 32
Hughes, Howard, 21–22

Humane Society of the United States, 32, 79, 163
Humphrey, Hubert, 39

Iceland, 3, 43
India, 86
Indonesia, 68
interdependence theory, 16–17
Interior Department, 4, 23, 47, 48, 156
internationalists (U.S.), 8–9, 39, 107–8, 114, 131. *See also* nationalists (U.S.)
Israel, 144

Jackson, Henry, 38, 131
Jagota, P. S., 144
Japan, 144
Javits, Jacob, 123–24
Johns Hopkins University School of Advanced International Studies, 8, 58
Jordan, Hamilton, 74

Katz, Ronald, 92, 94
Kaufman, Alan, 97
Keeney, David, 12, 109, 158
Kennecott Copper Corporation, 12, 21, 109, 123
Kennedy, John F., 11
Kent, George, 61
Kenya, 89, 93
Keohane, Robert O., 16–17
Kimball, Lee, 7, 8, 15, 78, 82, 95, 164, 165, 166; advice to NGOs from, 162–63; answers negotiators' questions, 83–84; article in *Neptune* (1977), 87; on assisting delegates from developing nations, 70; attends signing in Jamaica, 147; background and personal traits of, 27–28; begins work with Neptune Group, 56–57; and belief that she worked for CIA, 14; on delegates from developing nations, 4; as executive director of Citizens for Ocean Law, 125; and journalists, 35, 157; knowledge of law-of-the-sea issues of, 11; and Methodist project during Carter years, 74; Methodists unable to hire, 63; and Moon Treaty, 124; as Neptune Group member, 28–29; on Neptune Group's contributions to UNCLOS III, 150–51; on Neptune Group's goals, 10; at 1975 Geneva session, 58–59; at 1978 Geneva session, 91–93; at 1980 sessions, 100–101; organizes 1979 programs, 97; organizes seminar in April 1977, 86; protreaty activities in 1981–83 of, 130, 133–36, 137–40, 143–46; and Leigh Ratiner, 109; on Reagan's change of policy, 130; on seabed mining bills, 112, 113, 115–17; on types of Neptune Group programs, 66; on U.S. decision not to attend Preparatory Commission meetings, 145. *See also* Neptune Group; Ocean Education Project; United Methodist Law of the Sea Project
Kirthisingha, Pali, 97
Kissinger, Henry, 49, 64, 67, 85
Kitsos, Tom, 109
Koburger, Charles, 67
Koh, Tommy T.B., 68, 72, 85, 87, 120, 121, 143, 156, 165; asks Neptune Group to arrange programs in 1979, 97; asks Neptune Group to set up meetings with outside experts, 85–86, 90; assesses Neptune Group's contributions, 151, 158; calls for vote on treaty, 144; criticizes Elliot Richardson's support for seabed mining legislation, 99; elected president of UNCLOS III, 130; with Alexander Haig, 142; heads negotiating group 2, 89;

and journalists, 82; on Lee
Kimball, 29; at Levering's orchard,
98–99; on John Logue, 33; and
Jim Magee, 102, 103; meets
Neptune Group staff in Caracas in
1974, 54; and MIT econometrics
model, 93–95; on Neptune
Group's seminars, 104; 1982
statement on difficulties facing the
negotiators of, 139; praises work
of Neptune Group at signing in
Jamaica, xviii, 148; on seabed
mining, 92; works with J. T.
Nyhart on financial issues, 91
Kronmiller, Theodore, 129, 133–35,
137, 142

Lake, Anthony, 74
Lake Mohonk Mountain House, 86
Latin America, 3, 43
law of the sea (traditional), 23
League of Nations, 9
League of Women Voters, 24
Lee, Roy, 152
Lehman, John, 129
Leipziger, Dan, 67
Leshaw, Dale, 93
Levering, Miriam, 7, 164, 165–66;
answers negotiators' questions, 83–84, 96; attends preparatory
conference in Geneva, 51–53;
attends signing in Jamaica, 147;
background and personal traits,
xviii–xix, 9, 13–16; becomes ill at
Geneva session, 92–93; begins
work on law of the sea, 20–21;
interrupts Northcutt Ely in 1978
meeting, 111; Lee Kimball on, 82;
on Tommy Koh, 100; and Koh's
opposition to seabed mining bills,
121; and letter from delegate from
Mauritius, 150; at 1975 Geneva
session, 57–62; at 1982 session,
143, 144; and Public Advisory
Committee, 47–48; on Leigh
Ratiner, 110; on Reagan's stance,
129–30, 137; relations with
journalists of, 157; serves on
board of Citizens for Ocean Law,
125; on UNCLOS III, 14; on
what Neptune Group learned,
161–62; urges selection of Elliot
Richardson as U.S. negotiator, 74;
See also Neptune; Neptune Group;
Ocean Education Project; Quakers;
Sea Breezes
Levering, Sam, 7, 8, 96, 160, 165–66; activities at 1974 Caracas
session, 53–54; and George
Aldrich, 76–77; assessment of
work with Congress, 157–59;
attends signing in Jamaica, 147;
background and personal traits of,
9–13; begins work on law of the
sea, 20–21; and congressional
resolution supporting UNCLOS
III, 41–42; congressional testimony of, 38–39; cuts back U.S.
Committee's lobbying efforts in
1980, 126; declines appointment
to U.S. delegation, 57; heads
environmental subcommittee, 81;
invites John Logue to testify
before Congress, 32; on Jim
Magee's work, 102; at 1978
brainstorming conference, 90;
opposes mandatory transfers of
technology, 77; Charles Percy on,
xvii–xviii; and Public Advisory
Committee, 47, 49; on Leigh
Ratiner, 110; on Reagan's stance,
137; and Elliot Richardson, 74,
122; and seabed mining bills,
112–15, 118; seeks compromise
on seabed mining, 57; and
Treasury Department officials, 48;
and 200-mile fishing bills, 43–46;
Washington Post publishes letters of,
36, 115; works with State Department officials, 46–47. *See also*
Neptune Group; seabed mining;
seabed mining bills; United States
Committee for the Oceans

Lewis, A. Barton, 166; and Citizens for Ocean Law, 125; and Alexander Haig, 141–42; and law of the sea, 19–20; on Miriam Levering, 15; at 1982 session, 139; and Reagan's stance, 138; as world federalist, 9
L-5 Society, 123
Library of Congress, 40
Lilly Foundation, 25
Lindsey, Carol, 58
Lipton, Charles, 86
Lockheed Corporation, xx, 21, 48
Logue, John, 32–33, 38, 41, 55, 61, 160

Maechling, Charles Jr., 125
Magee, Jim, 100, 101–3, 131
Malone, James, 129, 131, 133–35, 136, 137
manganese nodules, 21. *See also* seabed mining
Manguson, Warren, 44, 45
Maryknolls, 41
Mason, Anne, 42
Massachusetts Institute of Technology (MIT), 85, 86. *See also* MIT/Harvard econometric model
Mauritius, 150
Maw, Carlyle, 7, 49
McClements, Robert, Jr., 138
McColl, Richard, 12
McIntyre, Stuart, 23
McKelvey, Vincent, 23
McLean, John, 44, 46, 113
Meese, Edwin, 132, 142
Members of Congress for Peace Through Law, 8, 163
Metcalf, Lee, 37
Methodist Church. *See* Hamlin, Joyce; Kimball, Lee; United Methodist Church; United Methodist Church, Women's Division; United Methodist Law of the Sea Project; United Methodist UN program; Weaver, Barbara

Mexico, 85
Miles, Edward L., 29, 70
mining company representatives, 94, 96, 108–9, 126, 130, 158
Minneapolis Star and Tribune, 38
MIT/Harvard econometric model, 90–91, 93–95, 152
Mobil Oil Corporation, 67
Mohonk Mountain House, 67
Moon Treaty, 109, 110, 123–25
Moore, John Norton, 44, 57, 77, 117
Morgenthau, Hans, 11
Mott, Jessica, 68
Murphy, Cecily, 139, 147, 157
Murphy, John, 109, 112
Murphy-Breaux seabed mining bill, 112–17

Nandan, Satya, 104, 152
National Association of Manufacturers, 37
National Council of Churches, 41
National Education Association, 24–25
nationalists (U.S.), 106–8, 112, 114, 124–25, 146, 159. *See also* internationalists (U.S.)
National Journal, 108
National Parks and Conservation Association, 79
National Public Radio, 61, 69, 157
National Science Teachers Association, 24
Neptune, 7, 14, 26, 28, 93, 104; decision to publish fewer issues, 63; initial issues, 57–61; Lee Kimball and, 135–36, 140; Edwin Miles on, 70; representative topics (1976–80), 68, 83, 87, 96; Rene Wadlow and, 57
Neptune Group, xviii–xix, 7–10, 15; assessments of, 149–59; and board of Citizens for Ocean Law, 125; and delegates from developing nations, 70; environmental

concerns of, 36–37; funding for, 25; importance of women in, 82; lessons for other NGOs from, 159–64; John Logue and Elisabeth Mann Borgese and 32–34; at 1974 Caracas session, 53–55; in late 1974 and early 1975, 56–57; at 1975 Geneva session, 57–62; and 1976 negotiations, 65–69; and protecting whales, 79; protreaty activities during early Reagan years, 130–48; relations with journalists of, 68–69, 82, 130, 140, 157; Elliot Richardson and Tommy Koh and, 72; and seabed mining legislation, 36–42, 106–27; and 200-mile fishing bills, 43–46; and UNCLOS III during Carter years, 81–105. *See also* Kimball, Lee; Levering, Miriam; Levering, Sam; Ocean Education Project; United Methodist Law of the Sea Project; United States Committee for the Oceans; Weaver, Barbara
Netherlands, 86, 143
news media, xix, 157; and Methodist project, 27; at late 1970s sessions, 82; and 1974 Caracas session, 53, 54–55; at 1975 Geneva session, 61–62; at 1976 New York session, 68–69; at 1981 and 1982 New York sessions, 130, 140, 141. *See also* Neptune Group; United Methodist Law of the Sea Project; *American newspapers and magazines by name*
Newsweek, 69
New York Times, 69, 141
Nixon, Richard, 45, 49
Nixon administration, 49, 73
Njenga, Frank, 89, 93
Nordquist, Myron, 12, 23, 40, 42, 44, 46, 47
Norway, 52
Nye, Joseph S., 16–17
Nyhart, J. D., 85, 86, 90–91, 93

Ocean Education Project, 7, 107; goals of, 10; and Methodist project (1982); origins of, 24; Washington operations phased out in 1983, 148. *See also* American Freedom Association; Levering, Miriam; Neptune Group; United States Committee for the Oceans
ocean fishing issues, 43–46
Oceanic Society, 58
Oceans: Time for Decision, The, 56
Ogley, Roderick, 58, 60, 63
Olmstead, Cecil, 125
Orr, Jim, 14, 58, 60
Oxman, Bernard, 20, 162, 164; and environmental movement, 81; and Sam Levering, 12–13, 23, 27; and Neptune Group, 141, 152

Pacem in Maribus, 33
Pal, Mati, 102, 103
Pan, 57
Pardo, Arvid, 33, 57–58
Park, Choon Ho, 68
Paterson, Arthur, 15, 70, 139, 166; contributions to Neptune Group of, 25–26, 53–55, 63; and journalists, 68–69, 157; on Lee Kimball and Barbara Weaver, 29, 30; on Neptune Group's work during Carter years, 104; on role of women in Neptune Group, 82; shows concern for whales, 79, 81
peace movement (U.S.), 9. *See also* American Freedom Association; American Friends Service Committee; Center for Concern; Friends Committee on National Legislation; internationalists (U.S.); Members of Congress for Peace Through Law; Quakers; Stanley Foundation; United Methodist Church; United Nations Association; United World Federalists
Pell, Claiborne, 39, 42, 106, 117–18
Percy, Charles, xvii–xviii, 118

Perlmutter, Howard, 97, 98
Peru, 45, 92
Peterson, Russell, 23
Peterson, Stephen, 67
Philadelphia Inquirer, 69
Philippines, 68
Pinto, Christopher, 64
Platt, Stanley, 165–66
Pontecorvo, Guilio, 3
Preparatory Commission, 155
Public Advisory Committee to the Joint Interagency Task Force on Law of the Sea, 73, 160; Lee Kimball's work with, 133, 137, 145; Sam Levering's work with, 12–13, 46, 75, 133, 145; participation of environmentalists in, 32, 80–81. *See also* Commerce Department; Defense Department; Interior Department; State Department; Treasury Department
public opinion (U.S.), xix

Quaker House (Geneva), 8, 60, 82, 102
Quaker House (New York), 24, 51, 65, 68, 69
Quakers (Society of Friends), 9; assistance to the Neptune Group of, 58, 60, 67, 69, 82, 103, 104; Sam and Miriam Levering's work with, 11, 13. *See also* American Friends Service Committee; Friends Committee on National Legislation; Friends United Meeting; Friends World Committee for Consultation; Mohonk Mountain House; Quaker House (Geneva); Quaker House (New York); Quaker UN Program
Quaker UN Program, 8, 65

Ralston Purina Company, 44
Ratiner, Leigh, 48, 73, 154–55, 165; helps to write 1970 draft treaty, 20; and Sam Levering, 23, 47; on John Logue, 33; and Moon Treaty, 123–24; opposes U.S. policy during Carter years, 77, 78, 109–10, 112, 114, 115; praises Neptune Group's work at UNCLOS III, 70, 152 and seabed mining, 64; as U.S. negotiator under Reagan, 129, 132, 139–42, 143; writes anonymous article for *Neptune,* 60–61
Reagan, Ronald, 45, 125, 132, 139, 142, 146, 164
Reagan administration, 125, 128–46, 159
Republican Party, 107
Richardson, Elliot, 20, 109, 115, 118, 129, 155, 156, 158, 159, 165; as chief U.S. negotiator, 96, 103; and delegates from developing countries, 104; and environmentalists, 32, 78–81; and formation of protreaty organization, 125; and Tommy Koh, 99; on Leverings, 74–75; and Neptune Group at UNCLOS III, 82, 90; on Neptune Group, xix–xx, 152; Reagan administration and, 130; recommended to head U.S. delegation, 74; and seabed mining legislation, 91–95, 112, 113, 117, 119, 121; at signing in Jamaica, 147. *See also* Carter administration; State Department
Riessen, Mary Beth, 26, 27, 41, 82, 166
Rishel, Brad, 44
Rishel, Hannah, 44
Rockefeller Foundation, 65
Roman Catholic Church, 41. *See also* Center for Concern; Riessen, Mary Beth
Rusk, Dean, 42, 74
Russia. *See* Union of Soviet Socialist Republics

Safire, William, 107
Samuelson, Robert J., 108

Sanger, Clyde, 3, 150
Sargent, Bill, 53
Scharlin-Rambach, Patricia, 24, 51
Schmidt, Markus G., 76, 150, 158–59
Schnurr, Eleanor, 24
seabed mining: as issue at UNCLOS III, 5–6, 64–65, 84–87, 93–95, 135–36; in U.S. politics and policy making, 77–78
seabed mining bills, 20–21, 36–42, 106–27. *See also* Congress; Levering, Sam
seabed mining companies. *See* mining company representatives
Sea Breezes, 47
Sebenius, James K., 91, 93, 94, 97, 150, 152
Sedco Corporation, 97
Seiberling, John, 42
Senate Commerce Committee, 44
Senate Finance Committee, 118
Senate Foreign Relations Committee, 106–7, 109, 118, 122
Senate Interior Committee, 41
Shell Oil Corporation, xx
Short, Ken, 122
Shultz, George, 48, 49
Sierra Club, 18, 24, 25, 32, 41; impact on UNCLOS III, xx, 79–81; and Neptune Group organizations, 8; testimony before Congress, 38; at UNCLOS III, 51, 53, 80–81. *See also* Clusen, Charles; environmental issues and organizations; Sargent, Bill; Scharlin-Rambach, Patricia; Yurchyshyn, Anita
Simon, William, 48
Singapore, 54, 68, 85
Smiley, Keith, 67
Smith, Anthony Wayne, 79
Smith, David N., 85, 86
Smith, Eleanor, 29, 102, 135, 150, 160, 161
Smith, John T., 76
Snyder, Ed, 19, 20, 42
Sobhy, Samir, 85

social movement theory, 17–18
Sohn, Louis, 20, 23, 41, 63
Sondaal, Hans, 86
Soundings, 26, 28, 70–71
Sound Ocean Systems corporation, 98
Sri Lanka, 91
Stanley, Maxwell, 66
Stanley Foundation, 8, 65, 66, 90
State Department, 40, 111, 156; on fishing legislation, 44, 45, 46; Sam Levering and, xviii–xix, 47; and ocean law, 4, 23; and Treasury Department, 49
Stelle, William, 165
Stennis, John, 42
Stevens, Ted, 146
Stevenson, John R., 12, 40, 42, 49
Studds, Gerry, 45
Sullivan, Leonor K., 44, 45
Sun Companies, 138
Swing, John Temple, 125, 152

Taft, George, 47, 81
Tanzania, 60
Teller, Edward, 11
Tenneco Corporation, 21, 22
Third UN Conference on Law of the Sea. *See* UNCLOS III
Third World. *See* developing nations; Group of 77 developing nations
Thomas, Christopher, 85
Thurmond, Strom, 42
Time, 69
Tipson, Fred, 107, 122
Train, Russell, 125
Transportation Department, 47
Treasury Department, 48–49, 156
Trinidad and Tobago, 85
Tsongas, Paul, 122
"tuna wars," 3, 43
Turkey, 144

UNCLOS I (First UN Conference on Law of the Sea), 3–4
UNCLOS II (Second UN Conference on Law of the Sea), 3–4

UNCLOS III (Third UN Conference on Law of the Sea); accomplishments of, 6–7; background and duration of, 3–6; developments following conference's conclusion, 6–7; 1974 session in Caracas, 53–55; 1975 session in Geneva, 57–62; 1976 sessions in New York, 63–70; 1977 session in New York, 75–76, 85–87; 1978 session in Geneva and New York, 89–95; 1979 session in Geneva and New York, 96–98; 1980 session in New York and Geneva, 100–103; 1981 session in New York and Geneva, 128–30, 135–36; 1982 session in New York, xvii, 141–44; signing session in Jamaica, xviii, 147–48; U.S. policy toward, 4–7; views of Western business and political leaders toward, 6. *See also* developed nations; developing nations; Kimball, Lee; Koh, Tommy; Neptune Group; Richardson, Elliot

Union of Soviet Socialist Republics (USSR), 79, 81; and restrictions on ocean research, 46; and seabed mining, 80, 102; and treaty expansionism in Third World, 107, 129. *See also* communist nations

Unitarian-Universalist Women, 24

United Methodist Church, 9, 104, 110; and liberal internationalist causes, 41, 42; Miriam Levering and, 13; UN church center, 13. *See also* Hamlin, Joyce; Neptune Group; United Methodist Church, Women's Division; United Methodist Law of the Sea Project; United Methodist UN program; Weaver, Barbara

United Methodist Church, Women's Division, 15, 26–27, 56

United Methodist Law of the Sea Project (UMLSP), 7, 44, 56–57, 125, 157; hires Arthur Paterson and publishes *Soundings,* 62–63; and Moon Treaty, 124; at 1975 Geneva session, 58,60; origins and activities of, 26, 28, 30; prayer service in New York, 87; on protecting the environment, 10; protreaty activities during early Reagan years, 130–31, 133–40, 143–47; public support for, 107; and seabed mining bills, 112, 115–17, 126; at UNCLOS III, 82; *See also* Hamlin, Joyce; Kimball, Lee; Paterson, Arthur; *Soundings;* Weaver; Barbara

United Methodist UN program, 26, 65, 82

United Nations, 9, 51, 79, 97. *See also* UNCLOS I, UNCLOS II, UNCLOS III

United Nations Association, 9, 24, 41

United Nations Center for Transnational Corporations, 86

United Nations Committee on the Peaceful Uses of Outer Space, 123

United Nations Conference on the Human Environment (1972), 30, 59

United Nations Conference on Trade and Development, 97

United Nations Legal Office, 152

United Nations Seabed Committee, 21

United Nations World Food Conference (1974), 30, 57

United States Coast Guard, 67

United States Committee for the Oceans, 7; and Congress, 41–42; origins of, 23–24; publishes *Sea Breezes,* 15; during Reagan years, 131; 200-mile fishing bills, 43–46. *See also* Levering, Sam; seabed mining; seabed mining bills; *Sea Breezes*

United States Geological Survey, 23, 67

United States Steel Corporation, 21, 109, 126

United Technologies Corporation, 124

United World Federalists, 9, 13, 18, 41, 107
U.S. draft convention on law of the sea (1970), 19–20

Vance, Cyrus, 73, 74, 123–24
Venezuela, 144
Vietnam, 67
Vietnam War, 11, 34, 107, 110

Wadlow, Rene, 57
Waelde, Thomas, 86
Wall Street Journal, 107
Warioba, Joseph, 60
War/Peace Institute, 32
Washington Post: advertisements opposing Moon Treaty in, 123, 124; and letter from Sam Levering (1978), 115; and Neptune Group members, 56, 69; on Neptune Group's stance (1974), 35–36; on Leigh Ratiner (1981), 140–41; on Third World delegates (1977), 108. *See also* news media
Weaver, Barbara, 7, 8, 9, 15, 56, 62, 63, 67, 82, 107, 139, 166; attends signing in Jamaica, 147; background and personal traits, 28; and mining bills, 112; and Neptune Group, 28, 30; praised by Charles Percy, xvii–xviii; and prayer service; works to form a broad-based protreaty organization, 125. *See also* United Methodist Law of the Sea Project
Weicker, Lowell, 79
Welling, Conrad, 48
Wells, Louis T., 66, 86, 98

Wharton School of Finance, 97
White, Carolyn, 58
William Penn House, 20, 24
Wilson, George C., 36
women (in Neptune Group), 82
women's rights movement (U.S.), 9
Woods Hole Oceanographic Institution, 125
World Affairs Council of Philadelphia, 8, 49, 93
World Association of World Federalists, 57
World Bank, 67, 85
world federalism, 33, 39, 165. *See also* internationalists (U.S.); United World Federalists; World Association of World Federalists
world-order organizations, 31, 32. *See also* American Freedom Association; Members of Congress for Peace Through Law; Neptune Group; Stanley Foundation; United Nations Association; United World Federalists; War/Peace Institute; World Order Research Institute
World Order Research Institute, 32
World Wildlife Fund, 125

Yu, Geoffrey, 54
Yurchyshyn, Anita, 32, 79, 80–81

Zaire, 39
Zuleta, Bernado, 147; on Neptune Group's assistance to UNCLOS III, 90, 102; on Neptune Group's contribution, 151–52; praises Neptune Group's work, 69, 95, 150